REBEL IN RADIO

REBEL IN RADIO

The Story of WQXR

by

ELLIOTT M. SANGER

KRAUS REPRINT

MILLWOOD, N.Y.

LIBRARY OF CONGRESS CATALOGING-IN-PUBLICATION DATA

SANGER, ELLIOTT M.

REBEL IN RADIO.

REPRINT. ORIGINALLY PUBLISHED: NEW YORK : HASTINGS HOUSE, 1973.

1. WQXR (RADIO STATION : NEW YORK, N.Y.) I. TITLE.

HE8698.S33 1986 384.54′53′097471 86-10511

ISBN 0-8115-0016-0

PRINTED IN THE UNITED STATES OF AMERICA

To ELEANOR
whose constant encouragement
helped WQXR over the rough spots

Contents

Introduction to the 50th Anniversary Edition

In 1956 when I was a 12th grade student at Syracuse, New York's, Nottingham High School, I first heard WQXR Radio via the WQXR Network's affiliate in Central New York, WDDS-FM.

At that time there were no all-news, no all-rock, no all-anything stations. Most radio stations offered a mix of pre-rock pop, such as Eddie Fisher, Kay Starr and Teresa Brewer, some news, and the few radio network shows that survived TV's coming of age.

Here, however, was something completely different. The enduring music of Mozart, Brahms, and Puccini—presented by voices that were at once warm but authoritative. Here were full-length concerts—live from Boston or New York—in real hi-fi. Here was news—radio news—from *The New York Times*, itself. What a revelation!

Now, 30 years later, there is no longer a WQXR Network, *per se*, and the station on WDDS's frequency broadcasts only rock music. Yet the radio stations that Elliott M. Sanger co-founded in 1936 and networked in 1956 have become, in 1986, the beacon on which scores of other broadcasters of quality music depend to set the standard for the industry.

This Anniversary edition of *Rebel in Radio* honors not only WQXR's founders, but also the hundreds of people who have put so much of their talent and effort into building its success. When Mr. Sanger wrote *Rebel in Radio* in 1973, the stations had just emerged from a severe period in their history. As Mr. Sanger describes, the F.C.C. had mandated that co-owned AM and FM stations in the same market substantially program different services—your FM could not duplicate your AM. For WQXR this was a blueprint for disaster since, at the time, station revenues were barely meeting the costs of supplying *one* classical music schedule over both channels. For six years the stations struggled with meeting the F.C.C. ruling. But in 1972, the F.C.C. granted a duplication waiver to co-owned AM/FM commercial classical music stations, including WQXR, and that ruling paved the way for the future of WQXR. And what a future we've had.

In the 13 years since *Rebel in Radio* was first published, WQXR has flourished as a New York Times Company subsidiary. Today, WQXR AM/FM are still America's #1 classical music stations. Over 700,000 people hear us every week, by far the largest audience for classical music radio. Advertising revenues have been climbing steadily for the past 10 years so that today WQXR's share of New York radio revenues is twice as high as our share of audience. Listener response—both to programming and advertising—is abundant, constant, and articulate.

I wonder if in 1936, in the depths of the Depression, Elliott Sanger and John V. L. Hogan could foresee the impact of their venture 50 years hence. Today, WQXR programming offers a daily listening staple for New York's most influential citizens, "from banking to Broadway," to paraphrase a sponsor's commercial. Our advertising roster includes the most important companies in America. Recently, for example, each of Fortune 500's top six companies sponsored at least one 26-week program series on WQXR. And now, WQXR can be heard all over the Western Hemisphere via satellite and cable hookups, carrying our programming to small cities in Louisiana, and to the Virgin Islands, as well as to 65th Street and Broadway.

To celebrate this 50th Anniversary year, we are rebroadcasting hundreds of moments of WQXR's history, which have been carefully preserved on tape and records. Through these segments, our listeners can appreciate the sounds and voices that gave WQXR so much of its character.

And through this book you may gain a better idea of just what it took to breathe life into WQXR and to keep the stations true to their missions of quality, integrity, and profitability.

For 700,000 of us, thank you Elliott M. Sanger—and thank you John V. L. Hogan—for your vision and your courage. Fifty years later, we are the beneficiaries of your legacy.

WARREN G. BODOW
PRESIDENT AND GENERAL MANAGER
WQXR AM AND FM

New York, April 1986

WQXR

Preface

Much of my adult life has been so inextricably woven with WQXR that this book may sound autobiographical. But that is not my intention. What I am trying to tell is how the idea of WQXR germinated, how it blazed a trail in the then wilderness of broadcasting and what effect it had upon the musical life of New York and its influence elsewhere.

Most of the credit for the idea of a station emphasizing the best in music must go to my partner in the early days, John Vincent Lawless Hogan. He was a man of science and an idealist. A fellow broadcaster once described him as "the most practical engineer dreamer I ever encountered." Much of what WQXR became was due to his abiding confidence in his tenets of better broadcasting.

This is not a technical book about radio. It is the portrait of a station which was and still is different and which despite great odds made a place for itself in the history of radio. I hope that the mistakes we made will warn others from doing likewise and that the correct assumptions we made will help those who may in the future wish to raise the standards of the industry. One must never forget that despite the aim for the ideal, the attainment of that goal must be economically feasible. It requires no miracle to operate a high-principled, intellectual, artistic, truthful station. But you have to be practical and think of where your support will come from.

Most of the material in this book is the result of notes in the diary I kept for more than thirty years. And where my notes or memory failed I was greatly helped by my wife Eleanor, who recalled many details which I hope will add to the reader's interest in and enjoyment of "REBEL IN RADIO."

ELLIOTT M. SANGER

New York, January 1973

The following prophetic words were written in 1888 in a very popular Utopian novel which fancifully looked back from the year 2000 to the world of Boston in the 1880's.

> It appears to me, that if we could have devised an arrangement for providing everybody with music in their homes, perfect in quality, unlimited in quantity, suited to every mood, beginning and ceasing at will, we should have considered the limit of human felicity already attained . . .
>
> I am sure I never could imagine how those among you who depended at all on music, managed to endure the old-fashioned system of providing it . . . Music worth hearing must have been, I suppose, wholly out of reach of the masses, and obtainable by the most favored only occasionally, at great trouble, prodigious expense, and then for brief periods, arbitrarily fixed by somebody else . . .

Looking Backward 2000-1887

by Edward Bellamy

1

The Scene

It was 1936. The United States was slowly emerging from the Great Depression; the world was more or less at peace although the saber-rattling of Hitler made ominous sounds. It was three years before President Franklin D. Roosevelt would stand before a crude television camera on the opening day of the New York World's Fair to be seen on only a handful of TV receivers strategically placed in the New York City area. It was the first time a President had been seen on television. Although few saw him, millions heard his speech on radio.

Radio was king. In the whole history of mankind there had never been a means of communication so rapid, so universal, so influential and so much a part of the everyday life of people. The radio set had become the domestic hearth around which the family gathered to be entertained and informed. Radio was a family activity which consumed many of its leisure hours. Some people had heard about television, but to most it still seemed to be in the realm of science-fiction.

Those of you who were not around at the time may find it difficult to picture the grip that radio had upon millions of people—perhaps even stronger than the hold that television has today.

There were magic names, dozens of them: George Burns, Gracie Allen, Edgar Bergen, Fred Allen, H. V. Kaltenborn, Raymond Gram Swing, Kate Smith, Rudy Vallee and many more who attracted millions of listeners to their programs. The great majority of Americans thought of these people as friends who were always welcome in their homes.

It was the era of "Amos 'n' Andy." When Calvin Coolidge was President, he left instructions that he was not to be disturbed by matters of state while they were on the air. Listening to the nightly adventures of these comedians became almost a family rite in the American home. The program was broadcast at seven o'clock Eastern Time, five nights a week for 15 minutes. In some towns, it kept so many people away from the movies that in order to hold their customers, the theaters stopped the picture at the scheduled time and piped in the radio program so that the audience would not miss an installment of the doings of Amos, Andy, the Kingfish, Lightnin' and Madam Queen.

As an instance of the great contrast between the entertainment world of 1936 and today, bear in mind that Amos, Andy and all the characters in this radio serial were blacks (played by white actors)—negroes of the old vaudeville image—typed as humorous, shiftless, irresponsible, loveable characters. They were just the types that black people today justifiably resent. In the 1970's it would be impossible, but in the 1930's, "Amos 'n' Andy" was a bond between the white and black communities—equally enjoyed by both.

High on the list of favorite programs of that era was the "soap opera" which filled so many broadcast hours every day. People lived through the sorrows and joys of the characters in serials such as "John's Other Wife", "Ma Perkins", "Mary Noble—Back Stage Wife", "Stella Dallas" and "Lorenzo Jones." It is doubtful whether today's TV dramas have the same vicarious appeal.

In 1936 radio was already such an important medium of communication that the Congress held an unusual night session so the maximum audience might hear President Roosevelt deliver a State of the Union message. King George V of England died that year, and all the world heard the tolling bells and the funeral service. It was a year of maiden voyages, and broadcasts came direct from the first trip of the "Queen Mary" and from the ill-fated zeppelin "Hindenburg." Probably, the most dramatic broadcast of the era, the abdication speech of King Edward VIII, was heard all over the world on December 11. That was the year WQXR was founded.

There certainly was plenty of programming available to the public at that time. Most people seemed to be satisfied; advertisers were finding that radio was one of the most effective mediums they had ever used; networks brought a great variety of programs to almost every corner of the country, and the networks and local stations were making money.

In the face of what was apparent saturation in this relatively new industry, a new radio idea was born. On a rainy night in the late Fall of 1935, two men were having dinner at the old Hotel Bossert overlooking New York Harbor from Brooklyn Heights. As they ate, they talked about radio programs and particularly about the need for more good music on the air.

One of them was a well-known communications engineer, John Vincent Lawless Hogan, a Yale graduate who had been in radio since the days of "wireless." He had contributed many inventions to the broadcast art, the best known being the "single dial control" which made it unnecessary for the listener to twiddle three or four knobs on his set in order to tune in a program.

Hogan had an experimental electronics laboratory on the top floor of a Ford agency garage in Long Island City. Among other projects he had been experimenting with mechanical television, and the sound accompanying his experiments was transmitted over his experimental radio station W2XR, the numeral 2 signifying that it was a noncommercial laboratory operation.

Because Jack Hogan liked good music, whenever W2XR was on the air he used phonograph records of the classics. People searching the air waves for relief from routine fare, happened upon beautiful music coming from a station they had never heard of. Without any urging, many of them wrote W2XR saying how much they enjoyed the music and expressed the hope that the program would be continued and expanded on a regular daily basis.

As far back as those experimental days, the station was attracting some attention. Ben Gross, the radio editor of the New York *Daily News* which had the largest circulation in the city, wrote this in his column:

> What chance has a small station of limited resources against the big ones with adequate staffs and fat bank accounts? Every chance in the world, provided those who run the station are fortunate enough to have an idea. Consider for example, tiny W2XR, a local studio, on the air but four hours a day—from 4 to 8 P.M. Rather than offer mediocre crooners and other poor talent, it specializes in recordings of the masterpieces in music . . . symphonies, concertos and the lighter classics. As a result it has built up a surprisingly big following. As the squire of a lordly estate in Westchester recently said to me: "Tuning in this station has become a habit. I'm always sure of something worthwhile." And if you ask me, in these few words of the suburban sage is the secret of audience appeal which so many program managers haven't learned yet.

It was the encouraging comments from the general public which these two men discussed during dinner and afterwards. Hogan showed some of the letters to the other man at the table, an old friend—the present writer. As far as radio was concerned, I was just a member of the listening public. I had owned a receiver from the very early days of the "cat's whisker" set. The technical miracle of how it worked was beyond me. I was a graduate of the School of Journalism, Columbia University, and my past experience had been chiefly in the fields of writing, advertising, publishing and merchandising. But I loved good music, though I never got much beyond playing scales on the piano. I went to concerts regularly and to the opera occasionally. Jack Hogan was no trained musician either, but he could play a few favorite pieces on the piano, and he loved to pluck out old tunes on a mandolin.

As the evening progressed, we began to discuss the possibility of making W2XR a radio station which would be self-supporting and possibly profitable on a policy stressing the best in music.

The talk went on for several hours, each of us weighing the difficulties of getting such a station started and keeping it alive. We agreed that there was enough run-of-the-mill programming offered by the 25 or more stations in the New York area so that there was no need for just another station. Our only chance would be a program formula which was entirely different from that of any other broadcaster—a station designed to serve a special rather than a general radio audience—and according to our thinking, it was the substantial potential audience of music-lovers.

Jack Hogan and I talked a lot about advertising. As radio had become more and more a mass medium, sponsors and stations allowed the sound and content of advertising to deteriorate. The ads had become noisy, long and "high pressure." Almost any kind of product was thought suitable for on-the-air plugging. Although part of the public resented the commercials, on the whole, listeners accepted them as a necessary evil and responded. The advertisers, of course, were happy.

Both of us realized that it would be difficult enough to convince advertisers that good music would attract a sufficient audience but if, in addition, we had to sell them on the idea of restrained, intelligent commercials, we knew our hurdles would be multiplied. We came to the conclusion that we had to "go the whole hog." We could not expect our prospective audience to listen to a singing jingle or a shouting pitchman sandwiched between Beethoven and Brahms. I do not think we anticipated how difficult this policy would be to maintain. If we had, we might never have gone ahead with the launching of WQXR. But we were sure that only a new kind of radio advertising must accompany a new kind of program. Proposed programs and advertising policies were sketched out as we talked that night.

After the first dinner in Brooklyn Heights, we met frequently over the next few weeks, and an agreement was made between Jack and me in February 1936 to form the Interstate Broadcasting Company which was a rather ambitious name for a radio station of 250-watts whose coverage was far from being interstate. The name suggests we had hopes for the future. The agreement provided that application would be made to the Federal Communications Commission for permission to transfer W2XR from the personal ownership of Hogan to a new corporation and to increase its power to 1000-watts. Under our new agreement, the voting majority of the stock of the new corporation was to remain with Hogan, and I put up $10,000 for a minority interest. That may not seem like a lot of money now, but in those latter days of the Great Depression, it meant about all the liquid capital I had. We both agreed that we were taking a big gamble, but we had faith in the idea.

Now that we had a corporation there had to be a place

for me to work so I moved into a corner of Hogan's engineering office at 41 Park Row in Manhattan. An interesting coincidence is that this building facing New York's City Hall had been the home of the *New York Times* before it moved in the early 1900's to Times Square. In 1936, we did not even dream that eight years later the *Times* would buy WQXR, and we would become a unit of that important newspaper.

Actual broadcasting was done from Hogan's laboratory in Long Island City. In a corner of it we rigged up some monk's cloth curtains to form a cubicle which became our "studio." Here, above the garage, the W2XR transmitter and the announcer-engineer combination worked in the makeshift studio. The building was not far from the Long Island City end of the Queensborough Bridge, and from the bridge, you could see the wooden pole and wires of our homemade antenna on the roof of the garage building.

The transmitter power of the station at that time was the minimum for any broadcaster—250-watts or one-quarter of a kilowatt, about half of what your electric toaster uses! The broadcast signal was just about strong enough to reach across the East River to midtown Manhattan and to parts of Queens and Brooklyn.

The newly formed company decided that the first money to be spent would be to raise the power of our transmitter and thereby increase our potential audience. In order to save money, a new 1000-watt transmitter was built by one of Hogan's laboratory engineers, Russell D. Valentine, who later became the station's Chief Engineer. We also bought a tall steel antenna pole which was erected on the roof of the laboratory. We were in business.

But there was one more big hurdle to jump. Our assignment on the dial was 1550 kilocycles (later shifted by the FCC to 1560) and, in those days, practically all radio receiver dials did not show any markings above 1500. Actually, many of the sets could receive above the 1500 mark, but people did not know that until they were told to try. And the cheaper receivers were actually dead above 1500. It was necessary to urge the listener to go to the end of the dial and try to hear us, and failing to obtain an adequate signal, to buy a more modern set. The reason receivers did not have markings above 1500 was because the broadcast frequencies originally ended at 1500, and it was only a few years earlier that the spectrum had been extended by the Federal Communications Commission to 1600 kilocycles.

Not only was our dial position a hindrance to getting a larger audience, but it also hurt our selling efforts. Often we lost an interested potential advertiser because his home radio did not go that high or his office boy told him he could not hear the station on his $16 set.

As I have mentioned, the original call letters of the station, W2XR, were for an experimental operation, and when we received a commercial license in December 1936, it became necessary to select other call letters. We decided on WQXR because over the air it would sound so much like W2XR that few listeners would be conscious of the change. Also, some people write the capital letter Q like the numeral 2, and so the transition would be minimal, both visually and orally. To this day, people write the station that they have been listeners since we were W2XR.

In the new company, Jack Hogan was president and I was vice-president. The latter title meant that I was to be chief salesman, program manager, station manager and supervisor of a staff consisting of a time-salesman and a girl secretary. This young lady, by the way, shared our only telephone with me, and she came down with the mumps the first few weeks she was with us. The fact that I did not catch it from her is one of the proofs that Heaven was watching over W2XR.

2

Program Policy

At our very first meeting, Jack Hogan and I decided to operate the proposed station more like a newspaper or magazine than a radio station. The networks and stations had become, to a large extent, just vehicles for carrying programs created and produced by sponsors and advertising agencies. If we wanted a higher cultural level for our station, we could not allow outsiders to supply the material. Following any other policy would prevent the creation of the kind of station we dreamed of. Therefore, we would "edit" our station, supplying the program and strictly controlling the advertising. The sponsor would associate his product or service with the content of the station's schedule. This basic rule meant that the sponsor would offer his advertising as he would to any well-run publication without control of the material in the adjacent columns of printed matter.

Our policy was designed to prevent the sponsor from bringing us a program which might meet his standards but not ours. In these many years, WQXR has accepted very few musical programs created outside our staff and those only when we believed that they fit into our program scheme.

Once in a while, yielding to the urgent need for revenue, we accepted programs which we hoped were appropriate but which later proved not to be. They were taken off the air as soon as possible.

Good music was to be the *raîson d'etre* of the station, but we did not intend to be a station appealing only to professional musicians or the musically trained. We wanted to reach the many thousands who liked good music but who were not being served by broadcasting as it then was.

At the beginning, we had a simple yardstick though it may seem unscientific and dictatorial. As I have said, neither Jack nor I were musicians or performers—we loved good music for the pleasure it gave us and we wanted to provide that same enjoyment to others. The criterion was simple: If we liked a composition, it went on; if we didn't, it was taboo.

This was a pragmatic policy to start with, but as the audience grew in size and in musical sophistication, we broadened the repertoire to include many works which we personally did not enjoy or even understand. When our programming was later in the hands of more musically trained people, we often had to restrain them from being too esoteric in their choices.

Several years later in applying to the FCC for a further increase in power, we defined our general program policy in six basic principles. Perhaps quoting them best presents our point of view on how our station was programmed:

1. Make every program either educational, cultural, informative or interesting (or a combination of those features).
2. Never tell the audience that the program is educational or cultural.
3. Never let a sponsored program (commercial) fall below the level required of a comparable sustaining program.
4. Put the emphasis on good music (of any kind) and avoid trashy "popular" music heard on many other stations.

5. Address all programs to people of intelligence and appreciation; never "talk down" to the audience and never underestimate its education or culture.
6. Always be sensitive to audience reaction and promptly modify any program to which the listeners make reasonable objection.

This sort of thinking was unusual in those days and is a platform that might improve today's radio and television. A few years after we started in business, Henry F. Pringle wrote an article in *Harper's* Magazine titled "WQXR: Quality On The Air." In it, he quoted Jack Hogan as saying, "We assume the radio audience to be an intelligent and cultured person" and then Pringle commented "It is an assumption which would qualify Mr. Hogan for a lunatic asylum in the minds of nearly all other radio company officials."

Other magazines ran articles as they discovered WQXR, and these created increased interest in what we were trying to do. One of the most effective was written for the *Saturday Review of Literature* by the well-known book publisher, M. Lincoln Schuster. His reaction to the station is summed up in this brief excerpt:

> I belong, then, to a goodly fellowship—the invisible legion of listeners to whom station WQXR is more than a broadcasting company, more than a miraculous and never-ending source of the world's best music, but almost literally a habit, a sanctuary and a way of life . . . If Keats were alive today, he would write a sonnet on first tuning in to station WQXR.

Our definition of good music was very broad. We did not intend to limit ourselves to the music of Beethoven, Brahms and Mozart or to Bach fugues. We always included in our definition the music of Gershwin, Rodgers, Porter, Kern, Berlin, Schwartz and others among the contemporary composers for the Broadway stage as well as the much-loved Viennese and other tuneful European operettas and, of course, Gilbert and Sullivan. We believed, and this was later confirmed by experience, that our audience could love Mozart but still enjoy Cole Porter, provided it was programmed at the proper time and in the proper environment. You can enjoy both, but you cannot put them in the same program segment.

There is a special problem in programming good music. An individual's taste changes over the years. Perhaps a listener begins to like good music upon hearing the overture to "William Tell" as the stirring signature of the popular radio program of the day, "The Lone Ranger." His next step might be Tchaikovsky's "1812 Overture." Then perhaps he next enjoys some favorite opera arias and later a symphony or a complete opera. It takes time for the novice in music to enjoy Mozart or Bach or even Beethoven. And progress is made in another direction also. From shorter pieces and the more tuneful symphonies, people move on to an understanding of the more sophisticated. Our experience showed that fully orchestrated melodic symphonies are enjoyed before the less grandiose compositions of Mozart and Haydn. As for trios, quartets and other chamber music, that enjoyment usually comes later. This fact made life more difficult for our program staff because once you appreciate chamber music, you never seem to get enough of it. The enthusiasts were always asking for more and more chamber works, and they are only a small but vociferous part of the audience. We found that out very soon by comparing the size of the audience for this music with the number who preferred symphonies, concertos and operas.

Another difficulty which the outsider may not realize is that this development in musical taste is cyclical. For example, a listener who had unconsciously advanced musically from the "Blue Danube" of Strauss to Ravel's "Valses Nobles et Sentimentales" may complain that we play too many Strauss waltzes. But he does not realize that in the five years or so that he has been listening, others have joined the audience who may not yet be ready for Ravel. The longer a classical music station is operating, the more complicated this problem becomes. In 30 years, one has at least two distinct generations of musical taste and a half-dozen or more levels of musical sophistication. To some the station became "low brow" while to others it was "high brow."

A graphic picture of how musical taste shifts over a period of years may be seen in a comparison of the results of two questionnaires WQXR sent to a large number of listeners at a 14-year interval. They were asked to list their favorite operas and favorite symphonies in the order of preference. We tabulated the top 30 in each category, but a list of the top 10 is sufficient to show what happened:

OPERAS

1962 Ranking	*Composer*	*1948 Ranking*
1. La Boheme	Puccini	6
2. Carmen	Bizet	1
3. Aida	Verdi	5
4. La Traviata	Verdi	3
5. Tosca	Puccini	18
6. Madama Butterfly	Puccini	10
7. Don Giovanni	Mozart	2
8. Rigoletto	Verdi	16
9. Tristan und Isolde	Wagner	4
10. La Nozze di Figaro	Mozart	9

The most startling change, as you will note, is the big jump in the popularity of the Puccini operas, the high rank of Carmen over the years and the stability of the love for Verdi.

SYMPHONIES

1962 Ranking	1948 Ranking
1. Beethoven No. 9	2
2. Beethoven No. 5	1
3. Beethoven No. 3	5
4. Beethoven No. 6	7
5. Beethoven No. 7	8
6. Tchaikovsky No. 6	4
7. Brahms No. 1	3
8. Dvorak No. 5	13
9. Brahms No. 4	9
10. Tchaikovsky No. 5	10

The obvious conclusion from this, of course, is the dominance of Beethoven. Dvorak was the only newcomer in the top 10.

We realized early in our history that we could not please everyone and to try to do so would drive us to the psychiatrist's couch. We explained this dilemma to people as best we could by saying, "We know we cannot please 100 per cent of the listeners 100 per cent of the time; our ambition is to please 90 per cent of the listeners 90 per cent of the time."

There were many other differences in taste which were hard to reconcile. The opera lovers vs. the symphonic buffs; the people who demanded more of the classics vs. the shouts for more contemporary and experimental music; those who wanted some talk about the music they were hearing and those who wanted their music straight; those who liked stirring music to get them started in the morning and those who could not tolerate anything at that time of day except slow soothing melody. These problems were ever with us, and in actual operation of a station, they defy solution.

In the beginning, life was very simple. W2XR was only on the air from 5 P.M. to 9 P.M. People were so glad to get

W2XR Feature Programs

1550 K. C. JUNE 1936 "At the End of the Dial"

DAILY FEATURES

5-6 p.m.—Cocktail Hour—Popular Music
6:00-6:45—Light Classics
6:45 (ex. Sun.)—Press Radio News
6:50-7:00—Light Classics
7-8 }
8-9 } See detailed program herewith

Monday, June 1

7 p.m. Gounod—Faust—Acts III and IV
8 p.m. Tschaikowsky—Symphony No. 4

Tuesday, June 2

7 p.m. Smetana—Quartette (From My Life)
8 p.m. Liszt—Symphonic Poem—Les Preludes

Wednesday, June 3

7 p.m. Sibelius—Symphony No. 4
8 p.m. Wagner—Symphonic Synthesis of Act III—Parsifal

Thursday, June 4

7 p.m. Tschaikowsky—Symphony No. 5
8 p.m. Bach—Chaconne in D Minor

Friday, June 5

7 p.m. Schumann—Trio in D Minor
8 p.m. Beethoven—Symphony No. 6

Saturday, June 6

7 p.m. Schumann—Symphony No. 7
8 p.m. Echoes from the Opera

Sunday, June 7

p.m. Grieg—Sonata in C Minor
8 p.m. Brahms—Symphony No. 1

Monday, June 8

7 p.m. Gounod—Faust—Acts IV and V
8 p.m. Liszt—Piano Concerto No. 2

Tuesday, June 9

7 p.m. Beethoven—Piano Trio No. 7
8 p.m. Bach—Toccata in F

Wednesday, June 10

7 p.m. Mozart—Symphony No. 39
8 p.m. Wagner—Excerpts from Das Rheingold

Thursday, June 11

7 p.m. Mozart—Quintette for Wind Instruments and Piano
8 p.m. Schubert—Unfinished Symphony

Friday, June 12

7 p.m. Haydn—Oxford Symphony
8 p.m. J. C. Bach—Sinfonia in B Flat Major

Saturday, June 13

7 p.m. Sibelius—Symphony No. 2
8 p.m. Echoes from the Opera

Sunday, June 14

7 p.m. Schumann—Sonata for Violin and Piano
8 p.m. Beethoven—Symphony No. 7

Monday, June 15

7 p.m. Verdi—Excerpts from Rigoletto
8 p.m. Mendelssohn—Violin Concerto

Tuesday, June 16

7 p.m. Berlioz—Fantastic Symphony in C Major
8 p.m. Mozart—Piano Concerto in A

Wednesday, June 17

7 p.m. Borodin—Symphony No. 2
8 p.m. Richard Strauss—Tod und Verklarung

Thursday, June 18

7 p.m. Brahms—Quartette in G Minor
8 p.m. Mozart—Symphony No. 40 in G Minor

Friday June 19

7 p.m. Beethoven—Eroica Symphony
8 p.m. Wagner—Excerpts from Die Walkure

Page from program guide, June 1936.

even this small ration of good music on their radios in 1936 that nobody complained. The earliest listing of our broadcasts which I have been able to find is for June 1936, the fifth month we were in business.

The more important music was broadcast from 7 to 9 in the evening, and the first week shows these were the featured compositions:

Monday	Gounod—Faust (Acts III and IV) Tchaikovsky—Symphony No. 4
Tuesday	Smetana—Quartet, orchestrated (from *My Life*) Liszt—Symphonic Poem—Les Preludes
Wednesday	Sibelius—Symphony No. 4 Wagner—Symphonic Synthesis of Act III of *Parsifal*
Thursday	Tchaikovsky—Symphony No. 5 Bach—Chaconne in D minor
Friday	Schumann—Trio in D minor Beethoven—Symphony No. 6
Saturday	Schumann—Symphony No. 7 Echoes from the Opera
Sunday	Grieg—Sonata in C minor Brahms—Symphony No. 1

Most people will agree that this was quite a feast of great music and there is no wonder that when dial explorers came across W2XR, usually by accident, they were astounded. By word of mouth, the "in" people of those days heard about it, listened and told their friends. The hours we chose for our broadcasts were deliberate, for we knew that when we went on the air at 5 P.M., people would be tuning all around the dial in a heretofore futile effort to escape from children's programs which more or less monopolized the air waves in the late afternoon and early evening.

Evidently it worked for we began to get mail in big batches. Here are some of the things people said:

Our dial is permanently set at WQXR . . . we don't bother much with any other programs.

I know millions will enjoy your work so keep everlastingly at it.

The broadcasting on your station is easily the finest on the air.

It is too bad you cannot hear us applaud.

This is my first letter to a broadcasting station and I've owned a radio for over ten years.

That was the message of hundreds of them, and it encouraged us to believe we were on the right track. I have always found that reading letters from listeners, whether they are for you or against you, is one of the fascinating things about running a radio station. The novelty of getting responses from all over and from all kinds of people intrigued us from the very start, and this fascination never wore off. One of the recurring phrases in the letters was "the dial of my radio is rusted to WQXR," and to this day, enthusiastic listeners often use the "rusted" simile.

As the mail increased, we had the problem of handling it with a very small office staff. When this situation showed no sign of abating (which fortunately it did not), I asked my wife Eleanor one day if she could spare a couple of hours to help us out. She came down to the office and started to work, and she remained for 23 years. For most of this time, she was the program director of the station and is as much responsible for the growth and accomplishments of WQXR as anyone. Many times when the going was rough and we thought we could never make our income equal our expenses, it was her enthusiasm and belief in our ultimate success that kept all of us determined to win out.

Of course we could not afford to hire any extra help so in our first studio we all had to pitch in and do the chores:

sweep out the place, shove the piano around, hang microphones and we even learned to handle control room equipment. This could be done then because we had no engineers' union to object. I remember that my son Kenneth, then about 13 years old, who was already ambitious to become an electrical engineer, used to come to the studio on Saturdays and help out doing odd jobs in the control room.

3

Advertising Policy

As you know, American broadcasting differs from the system in most other countries in that the costs of operation are met by income from advertising. Most nations have government-supported and government-operated stations which do not accept advertising. The largest and most successful of these is the British Broadcasting Corporation which is a corporation chartered by the British government. The millions of pounds needed annually for this service are raised by a fee of about $30 for color sets and $17 for black-and-white which every owner of a TV set pays and which is collected through the Post Office. A fee for radio sets was dropped in 1971. The fee paid by the public is termed an annual license, and it is illegal to have receivers without a license. This yields the BBC the equivalent of about $350 million a year—a tidy sum for providing a superior broadcast service.

The British method has worked satisfactorily over the years although there has been some movement recently toward a modified American commercial system for radio. There already is commercial television in Britain operated by privately owned corporations chartered by the government. Because of the nature of the British democracy and the British sense of fair play, there have been very few complaints of broadcasting being used to the advantage of the political party in power, but the same cannot be said for some other nations where the control of broadcasting has been a most powerful propaganda tool of the government.

I was in England during the closely contested election of 1970 and listened to the BBC regularly. It was quite evident to me that there was complete fairness in the allocation of time to the Labor, Conservative and Liberal candidates. The campaign in Britain is much shorter than in the United States. In 1970 I believe the Conservatives and Laborites each had five 10-minute broadcasts and three were assigned to the Liberals. Would that we could follow that example in our Presidential campaigns.

In government control at its best, which most people agree is the BBC, what has the public gained? It is true that non-commercial broadcasting can produce some cultural, documentary and other programs which are superior to the average run of American broadcasts, but many of the cultural efforts one encounters on the BBC do not attract a large audience in Britain judged by American standards, and would draw an even smaller one over here, for they are frequently dull or abstruse. But by operating several different networks, each with a distinct type of program, the BBC does manage to satisfy a variety of taste. To one who has listened to the BBC services at various times over a period of years, the conclusion must be that the average American listener would not like it as well as the fare he now receives, poor as it may be.

The purpose of this comment on the American vs. a foreign system is by way of stating the problem which faced Jack Hogan and me when we first made our plans for WQXR. At no time did we consider making it a non-commercial station. Neither of us had the financial resources to do this nor did we have the inclination to have it subsidized by a foundation or a group of wealthy patrons even if such support could be obtained. We wanted a business which would bring the public a better kind of radio and which, at the same time, would make a profit. Fortunately, we did not have a crystal ball available at that time, for it would have shown that it was going to take us seven years before we could make a small profit. If we had known, we might not have taken the risk.

But although we wanted a station supported by advertising, we were convinced that we could not have quality programs unless the advertising was in keeping with the atmosphere we hoped to maintain.

From the very start we wrote down some rules to guide us. They seemed revolutionary to the broadcasting and advertising fraternity, and that is why it took us so many years to escape from the red ink. These were the rules:

1. The station will accept no advertising of products which it believes to represent a bad value to the purchaser.
2. Even though a product or service offered for advertising on the station may be recognized to represent good value, it will not be advertised over that station if its character is such as to be obnoxious or offensive to WQXR listeners.
3. The station management will use every effort to prevail upon advertisers and agencies to rely upon messages which are factual and informative rather than exaggerated or blatant. The method of presentation must be in keeping with the quality of the broadcast program.

Looking back at the wording of these three points, they were rather pompous for a fledgling business to issue, but we meant it. The message was going to have to be the way we thought our listeners wanted it. Perhaps this anticipated the thoughts of Marshall McLuhan.

Although in these early days Jack and I had no idea that the station some day would be acquired by the *New York Times,* we believed that in its content and advertising it should try to match in radio the high journalistic and advertising standards of that newspaper. In fact, the original application of Hogan for a radio station stated that it would try to emulate the standards of the *Times.*

It is necessary to note here that there has been a decided relaxation of these standards in recent years. While I was in charge, I insisted that the quality of the advertising be maintained both as to the nature of the product or service and the method of presenting it on the air. After I retired in 1967, the management accepted many spots and programs which I would have vetoed. It is only fair to say, however, that the great increase in costs, particularly the labor union pressures for more and more money, plus the spiraling inflation of all costs, made it necessary to accept most of the advertising which was offered to the station. It was probably better in the long run to compromise on standards and thus keep WQXR alive than to be forced to give up the station and all it meant to the listening public.

You may be sure that advertisers did not flock to enroll under the banner of better radio. There was one sponsor who was with the station in its pioneer days, the Wanamaker Department Store then at 10th Street and Broadway, but it paid us with records instead of much-needed cash. That is how we started what has become one of the most complete record collections in the country. A few other advertisers came along, buying spot announcements at $5 each. It must be admitted that some of these pioneers were publisher friends of mine such as Simon and Schuster and Random House, and they were correct in thinking that WQXR listeners could read as well as listen. We also had orders in those first few months from the first of the movie "art theaters" in New York, the Plaza Theater, but there certainly was not enough to even pay our electric power bills. In July and August 1936, our gross income was less than $200 each month. It was not until September that we received our first contract from a real consumer product—

Martinson's Coffee. This was a high-quality coffee which sold at a premium price. It had a character all its own which was symbolized by its Rolls-Royce delivery auto. It was not long before we were known as "the Martinson Coffee station."

At about this time we decided to do a little promotion, and we bought a small advertisement in *Variety*, the show business weekly. The headline on the ad read, "WQXR—The Station for People Who Hate Radio!" This may have sounded smart, but it stirred up the advertising agencies who wrote letters to *Variety* asking by what right did this little two-by-four radio station put itself on a pedestal and criticize the great and booming radio industry which had brought the American public such fine entertainment. We certainly had our ears pinned back, but the objected-to slogan expressed exactly what we had in mind. It did not make the agencies more receptive to our solicitations for advertising.

Even when the few isolated spot announcements began to appear some listeners were surprised. We had gone through so many lean months that people thought that we did not intend to carry any advertising. Those who complained were informed promptly by us that we were a commercial station and that we hoped to get much more advertising because without it we could not continue to broadcast the kind of programs they told us were so welcome.

In addition to answering specific complaints by mail, Jack Hogan and I each went on the air every once in a while to explain to our listeners that the continuance of WQXR would depend on three-way cooperation among the audience, the advertiser and the station. The station would see to it that only reputable products and services were accepted; the advertiser had to cooperate in program quality combined with advertising copy that was aimed at an intelligent and cultured group; and the individual listener must make every effort to patronize our advertisers in order

A 1972 newspaper advertisement about the special nature of the "WQXR Family."

for them to know that they were getting results from their use of WQXR. We emphasized that if they really liked our programs and our approach, they would have to respond to our sponsors, otherwise WQXR could not stay on the air. These talks always brought an encouraging response, and they did much to knit together what we began to think of as the WQXR Family.

There are some people who instinctively hate advertising no matter how it is presented. This type of listener did not realize or even care that WQXR advertising was so strictly supervised. One of the interesting things we discovered was that among our most appreciative music-loving listeners were physicians, engineers, scientists and lawyers. They were also the most frequent critics of any and all advertising. Our only explanation of this is that those in the professions which by tradition and ethics bar advertising, seem to be the ones most resentful of it in any form.

To maintain our advertising standards has always been the station's most serious problem. As time went on and advertisers saw that WQXR covered a market which could not be reached by other radio stations, programs and announcements were offered us, many of which did not agree with our ideas of what we wanted and what we knew our listeners would respond to. When we offered to re-write or substitute our version of a commercial, we frequently came up against pride of authorship which vetoed any change from the inspired copy of the sponsor, writer or agency (and sometimes the sponsor's wife).

Fortunately, in many instances, we were able to convince the prospect that our way was worth a trial and, after a test, we were generally able to convince him that we knew best how to get results from our audience. One of the most successful examples of this cooperation between the station and the sponsor was with the ubiquitous, at the time, Pepsi-Cola "hits the spot" jingle with the "nickel, nickel, nickel" refrain. WQXR would not take it because we did not think that any singing commercials mixed with Brahms, Mozart and Shostakovich. This jingle became so popular that people were humming it, so we suggested to Pepsi-Cola that it could get the same sales result, maybe better, on WQXR by eliminating the words and voices. We experimented with playing the tune in many ways and on different instruments, and finally produced it on a celesta. The tinkling effect hit the spot. The sponsor used it on WQXR for many years and even got our permission to use it on other stations throughout the country.

WQXR eventually had to set up its own copy-writing department to modify commercials or create new ideas for many advertisers who were willing to try our approach. But there were hundreds of potential sponsors who would not budge from their own ideas. It is literally true that we rejected enough advertising to run another station profitably, but we stuck to our guns, not that we thought we had a mission but because we knew that once we changed our basic ideas we would have little left. WQXR was built on an idea, and we knew that if we down-graded our character, we would have nothing for the public and nothing for ourselves.

In the early years, Jack Hogan and I paid a great deal of attention to selling because without more business, we could not afford to operate. The man who led the battle for advertising through most of the years is Norman S. McGee who, when I became Chairman of the Board in August 1965, succeeded me as operating head of the station until he retired in 1968.

In WQXR's second year, a young lady in a large advertising agency asked me if I would see someone she knew who wanted to get into radio. Lots of people wanted jobs in radio at that time, and I said I would be glad to meet him. A few days later, in walked an attractive young fellow named McGee who said he had business experience, had been an actor out West and felt he could be useful to us. I told him the only opening would be as a salesman and that we could not afford to pay him a salary but would pay a commission on whatever he sold. He would have to think it over, he said, because it cost a lot to live in New York (even in those days).

As Norman McGee often told the story, a few days later he telephoned me and said he was willing to take a chance. According to him, I replied that I too had been thinking it over and did not know whether I wanted him. So I asked him to come in and see me again in a few days. Fortunately, for WQXR, for me and for him, I changed my mind again and hired him at no salary. But in a few months, he showed promise so we paid him $25 a week even though our resources were pretty slim. It was a good investment.

4

Sponsors

It was a red-letter day at WQXR because we expected a visit from representatives of the American Tobacco Company to audition a program which we had offered to them. As this was in the infancy of the station, we were about to have our first chance to get a large national advertiser as a sponsor. We expected some people from the advertising department of the big company, but we were overwhelmed when the president of the corporation, George Washington Hill, appeared with some of the other top officers.

In a sense this was natural, for Mr. Hill was the personality who dominated the advertising policy and who had made "Lucky Strike" cigarettes the leading brand of those days. Norman McGee and I were not only impressed but stunned at the great man's arrival. He marched into our rather simple quarters wearing a ten-gallon hat which he kept tilted on the back of his head as he put his feet up on a table and said he was ready to hear what we had for him.

Our program idea was called "The Treasury of Music" consisting of famous, tuneful, popular classics. And the "gimmick" we had included was to have in each broadcast one record of the voice of Caruso. We sat in the studio with

George Hill and anxiously watched his facial expression as the half-hour program unfolded. There was not much reaction noted on his face, but his eyebrows were raised a little when the Caruso record was played, and we explained that there would be one in each show. When the program was over, he removed his feet from the table, sat up, pounded his fist on the desk and said, "We'll make Caruso the Charlie McCarthy of this program!" In case you are too young to remember, Charlie McCarthy was a ventriloquist's dummy who sat on the lap of Edgar Bergen and amused the nation with his humor and caustic comments when radio was at its summit. It was one of the most successful radio programs for many years, and "Charlie" was a pet comedian of the whole country.

We were, of course, somewhat shocked by this comparison with Caruso, but who were we to question the advertising genius of George Washington Hill? He bought the program, and it was on the air with us a long time advertising Tareyton cigarettes successfully, long before the era of that same brand on TV with its "I'd rather fight than switch."

The history of WQXR is punctuated by strange encounters with a variety of sponsors and advertising agencies. One of our early thrills was the result of a telephone call from the J. Walter Thompson Company, probably the largest advertising agency then as it is now. The radio time-buyer wanted to see a salesman, for he thought he had a client for us. Norman McGee raced over to their office as fast as he could without appearing unduly excited (which he was), for this was out first inquiry from a large agency. With visions of a big order within his grasp, he talked about the then W2XR and inferred we could take care of any size order they might place. The next day they called again, and Norman went over to get the order. He came back to my office looking somewhat let down but he did have an order. It was for a single one-minute announcement to advertise the opening of the Hotel Ritz-Carlton Japanese Tea Garden Restaurant for the summer season. Singularly enough, that isolated spot attracted some customers to the Ritz.

To some advertisers the policy and content of WQXR did not make sense. They could not visualize anyone liking the kind of programs we had. But from the very start, there was one man who encouraged us by deed as well as word. He was the late Harry Scherman, founder and president of the Book-of-the-Month Club. His skill as an advertising man had made possible the widespread distribution of good books throughout the nation, and he thus realized that there was a cultured group which was ready for WQXR and which liked to read good books as well as listen to good music. He decided that our station could be an important factor in enhancing the cultural image of the Book Club. He sponsored excellent programs for many years. Some of these were recorded, some were live orchestras conducted by his son Thomas, who later became the conductor of the Little Orchestra Society.

Most of Harry Scherman's advertising copy was to explain to our audience how the Club worked and how it prevented people from failing to read the best books which were being published. His interest in WQXR was more than that of an ordinary advertiser. He felt that in addition to it being a good advertising medium for his company, WQXR represented a parallel cultural influence which should be encouraged. On and off, the Book-of-the-Month Club has been a sponsor for more than 30 years. Even today when I meet Harry's widow Bernadine, she always tells me how much she enjoys listening to WQXR. I remember with much gratitude how on one occasion when the station was struggling along, Harry Scherman loaned us a substantial amount to keep us alive.

Because WQXR has always been an "off-beat" station, it

has attracted some sponsors who one would not expect to advertise on radio. There have been a number of these, too many in fact to write about, but a few were so unusual that they deserve mention.

Who would ever imagine that the New York Yankees baseball club would sponsor a program of symphonic music? It did for a whole baseball season as an institutional project to reach people who might be interested in going to games but who were not willing to spend their time at the radio listening to a game. The program was on every afternoon at the same times games were being broadcast on other stations, and in the intervals between selections, baseball scores were announced. This series of programs attracted much publicity which was good for the Yankees and, of course, for WQXR. This was before the advent of TV which preempted the baseball scene. It was also before the Columbia Broadcasting System bought the Yankees.

Another unusual sponsor was the New York Stock Exchange. That conservative organization, always cautious of how it approached the general public, realized that WQXR listeners were people of better incomes and consequently interested in investments. The Exchange bought an hour every Friday night to present a symphonic program. In a brief two- or three-minute intermission, there was a message explaining the functions of the Exchange and the role it plays in the economic life of the country. This was one of our earliest experiences in financial advertising, which has since developed into an important classification of advertisers. The logic of the financial advertiser's selection of our station as an important medium in the field has been justified by the success so many of them have had in obtaining profitable response from the audience.

A few years after World War II we were approached by a young man named David Berger, representing the Association of German Broadcasters, who wanted to offer German music old and new. We accepted this unusual type of sponsor and Berger has presented this interesting broadcast weekly ever since.

Among the other oddities of sponsorship, I recall that one year at Christmas time, our seasonal broadcast of Handel's "Messiah," about three hours in length, was used by a chain of pork stores to convey its holiday greetings.

Sometimes we had to give much thought to accepting certain types of sponsors. One of these was the funeral home. There had been advertising of this nature on the air at other stations for years, but most of it was undignified, vulgar and musically inept. We were able to design a program of fine music, without funeral-like organ sounds, but great classics which were presented in good taste by the Universal Chapel in New York and which apparently was acceptable to our listeners.

We knew we could offer a program of this kind in keeping with the general atmosphere of the station. On the other hand, we also knew that there were other products and services which did not belong with us, no matter what we might do to try to make them acceptable. Among these were laxatives, corn cures, baby pants, remedies for athlete's foot, deodorants and similar personal products. Not that many of these items were not legitimate, safe and of good value, but we wanted no messages on the air which would jar the sensibilities of a person who was still in a mood of exhaltation after hearing a great masterwork.

Over the years, I believe our greatest problem has been advertisers and advertising. We had a staff which always could cope with program problems; we could always solve our labor problems, serious as they were at the time; our engineers could explore the newest developments in radio and keep up with technological advances, but the life blood of the station, the vital need for income from advertising, was with us day in and day out.

Our program and advertising policies forced us constantly to walk a tightrope, balancing in a precarious spot between art and commerce. The temptation to make compromises often endangered our footing on the high wire, but like a good circus performer, we generally could keep our balance and not endanger the valuable neck of the station. But it was not easy, particularly in the first five years when it was touch and go.

As everyone knows, radio and television time is generally bought by sponsors according to "ratings"—that mysterious yard-stick which is sometimes right and frequently wrong. The difficulty of sampling adequately the audience of WQXR made the usual methods of audience surveys unfair to us. The reason for this is simple: Because of our programming, the potential audience could never be as great as the audience for more popular mass broadcasting, and so when rating services went into the field to count listeners or to make telephone inquiries, the odds were high against finding a WQXR listener rather than a listener to one of the mass-appeal stations. The odds were even much greater against us when lower income areas were being canvassed.

Despite this handicap, WQXR had to face the realities of the advertising world. The advertising agencies want to buy according to the rating books. When they buy newspaper or magazine space for their clients' advertisements, they have certified circulation figures to back their judgment. In buying radio and TV, they want something similar, even though they too question the infallibility of ratings.

The rating services must necessarily rely on a sample of the total audience and project their figures from the findings in the sample. The cost of sampling the nation or even a local community comes high and necessarily keeps the sample small. Yet the rating organizations insist that the sample is selected so carefully that it is accurate. This conclusion I and many others have always doubted.

I remember hearing a talk by the head of one of the most reliable rating organizations. He recalled that at a cocktail party, a lady had come up to him and said, "How is it that with all these surveys going on, I have never been interviewed about my listening habits?" In reply, the executive said, "Your chance of being interviewed, madam, is just about the same as your chance of being struck by lightning." "Oh, but I *have* been struck by lightning!" was her quick rejoinder.

As the years went by, WQXR began to show increases in the ratings, but at no time did they indicate the real size of our audience which program response and other tests indicated. We realized from the start that we would have to prove that our audience, whatever its size, consisted of people who could not be reached by the regular run of radio programming. From the mail we were receiving, almost from the day we went on the air, we could tell that these were people of intelligence, culture and above-average in living standards and income. We used to think of this simply as a method of proving the quality of the listeners, but nowadays this type of survey has a more impressive label as a "demographic" analysis.

It is these demographic studies which enabled the station to attract the right kind of advertiser, for they proved in comparison to similar studies of competing stations that WQXR gave the advertiser a market high in income, a much larger incidence of college and post-graduate education, a far greater proportion of higher-income managerial jobs, a greater concentration of homes in better neighborhoods, more involved in financial affairs and investing, and a far above-average activity in travel in the United States and abroad on business and pleasure. Another significant result of these studies is that the audience is essentially adult and has very few listeners below the age of 18. This avoids much waste for those advertisers who are not interested in the teenage purchaser. Taken as a whole, it indicates that the station succeeded in attracting a special audience for itself, and even if it is not the largest, it is one of the most productive because of its purchasing power and its strategic importance in influencing others.

All this may seem like a "plug" for the station as an advertising medium. That is not the intention. An important part of the story of WQXR is its special audience, without which the station would have disappeared years ago.

"Music in the best sense does not require novelty: nay, the older it is, and the more we are accustomed to it, the greater the effect."

Goethe

5

High-Fidelity

WHEN WE started in business our station identification on the air was "W2XR—the High Fidelity Station." This meant little to the general public, but it meant a great deal to Jack Hogan. He insisted that to hear sound faithfully reproduced, especially fine music, existing radio and phonographs were inadequate. In explaining it to me who knew little about it, he pointed out that the average ear distinguishes sounds through a range of 60 to about 14,000 cycles per second. The best designed radio station of those days probably transmitted no frequencies above about 8,000 cps. He explained that the ordinary telephone was capable of transmitting only up to about 4,000 cps. and that is why our voices sound so different on the phone than when we speak face to face. He pointed out that when you listened to a symphony orchestra in Carnegie Hall, you heard the music through the full range of sound, but on a record or a broadcast, you heard about half the frequencies. It was comparable to looking at a great painting in good light or looking at it through a curtain of gauze.

Because each radio station was assigned a broadcast band that was only 10 kilocycles wide, it was impossible with equipment then available, to transmit the full range of sound frequencies in that width channel, so Hogan went to the FCC and asked for a double-width channel, 20 kilocycles. He obtained it on an experimental basis, and he then designed and built with Russ Valentine a transmitter which would deliver substantially all the listenable frequencies without distortion. Bear in mind that this was Hi-Fi on AM—long before FM (Frequency Modulation) was heard of, although FM's inventor, Edwin H. Armstrong, was experimenting with FM at the time. It was also before good recordings were available, most of them being accoustical records made with a megaphone. Electrical recording was just being introduced.

Leaving all technical explanations aside, all I knew was that WQXR's music sounded more real on the air than any other broadcast I had ever heard. Of course, you could not make a record sound any better than the frequencies that were on it, but when you heard "live" voices or instruments, it made a tremendous difference. At that time, Steinway sent a piano to our transmitter with a soloist to play it. When the experiment was over, the Steinway executives said that it was the first time they had heard a Steinway on a broadcast that sounded like a Steinway.

The logical marriage of high-fidelity with the best in music was the keystone of WQXR's technical approach to broadcasting. People did not care too much about the sound quality of a pop or jazz program, but they wanted classical music at its best. The listener's problem was that his receiver at home was not capable of lifelike reproduction. We were just coming out of the worst of the depression years, and most homes had inexpensive radios. For the most part, sets were made to meet low-price competition and had only a minimum of quality parts and inadequate speakers.

As part of our plan to familiarize listeners with the meaning of high fidelity, we put experiments on the air to demonstrate the range of human hearing. Certain evening time periods were set aside and announced in advance, and listeners were told that by tuning in, they could test the fidelity of their radios as well as their own hearing acuity.

Jack Hogan conducted these tests on the air himself. He would first explain that we would transmit a note of 1000 cycles per second, and the listener should set his volume at what he considered the normal level. Then our transmitter would start sending a tone, each announced before it came over the air: 800, 600, 400 and so on down the frequency range. Then he would start up from 1000 cycles and, by various steps, go up to 16,000 cycles per second. The listener was told to make a note at what point he could no longer hear any sound, both in the lower and upper ranges, and then write us to report what he had heard. Those who had poor receivers or were beyond middle age could generally hear only a limited range, and others were very proud to report how wonderful their sets were and how well their ears performed. After each of these demonstrations, we received hundreds of letters and post cards. The comments were most enthusiastic about the quality of our transmission. I recall one which said, "When your Mr. Hogan talks, he is so real that he spits right in my eye!"

Because so many of the replies indicated that most sets of that period were inadequate to reproduce sound faithfully, we tabulated the results of the tests and used them to try to convince manufacturers to make better receivers.

In response to the growing interest in better sound, which our tests proved, a few manufacturers began to improve their product, but for a long time there was no high-fidelity set available for the sound connoisseur unless he assembled the parts and built his own, as some did. We thought we saw an opportunity to remedy this scarcity and at the same time make some money for WQXR. Jack Hogan and some of his engineering associates designed a combination AM radio

Jack Hogan (right) and the author listening to the WQXR high-fidelity radio and phonograph in 1943.

receiver and phonograph which could reproduce without distortion the whole range of human hearing. The chassis was installed in an attractive walnut cabinet which was appropriate in a living room. When completed, it sounded wonderful, and we picked a radio manufacturer to make it in small quantities. Of course, we advertised it over WQXR which had the most concentrated market for it. The WQXR High Fidelity Receiver and Phonograph with an automatic record changer sold for $265, and in 1938 depression dollars that was a lot of money. This sales campaign was one of the first proofs that WQXR could sell high-priced products.

Those who bought it were highly pleased, and we heard from owners still enthusiastic about the sets 10 or 15 years later. However, we never reached the quantity sale we needed for the amount of time and attention we had to give to this project, and we stepped out of the field a few years later. By that time, some of the large manufacturers of radios were selling much improved sets.

Because of their mutual interest in perfecting sound, Jack Hogan had always worked closely with Major Edwin H. Armstrong, one of the great pioneers in radio who had made a fortune through his invention of the super-heterodyne circuit in the earlier days of radio. He had been experimenting for years on an entirely new method of radio transmission which he called "Frequency Modulation" (FM) as opposed to the original system of "Amplitude Modulation" (AM).

The great advantages of the Armstrong frequency modulation or FM system were that it made possible static-free broadcasting and perfected Hi-Fi. It transmitted sound so perfectly that it was difficult to distinguish it from the original source. It was logical that "The Major" would work with WQXR in making his invention available to the public. I can still see him in his shirt sleeves working around in our studios getting ready for the first FM experiment we were about to broadcast.

On July 18, 1939, Major Armstrong presented the first regularly scheduled program on FM radio with music which originated from WQXR. A specially installed high-fidelity telephone line carried the program from WQXR's studios to W2XMN, Major Armstrong's transmitter atop the Palisades at Alpine, New Jersey. The first two musical selections to be heard were Haydn's Symphony No. 100 and Tchaikovsky's "Francesca da Rimini." Very few people could hear this historic broadcast for, as far as we know, there were not more than 25 FM receivers in existence in the vicinity of New York. But those few who were able to receive the music heard real high-fidelity reproduction over the air for the first time and agreed that FM was a revolution in broadcast transmission.

WQXR had applied to the FCC for an FM station, and we received a license in 1939. It went on the air in November 1939 with the experimental call letters W2XQR and was the first FM station in New York City. The transmitter which was on temporary loan from Major Armstrong was installed atop the 54-story Chanin Building at 42nd Street and Lexington Avenue and WQXR remained at that site until it moved to the tower of the Empire State Building in December 1965.

No phase of broadcasting has had more ups and downs than FM. In the first few years following 1939, the big set manufacturers and networks whose stake was in AM did not take kindly to FM. Very few sets were available, but once a listener was able to buy or build one, he became a missionary for the new and much superior system. It began to take hold.

Then, just as we thought it was out of the woods, the FCC decided to move the entire FM band to another part of the broadcast spectrum. That ruling junked every set that was in use. It took several years before manufacturers made sets for the new band and before once-burned listeners ventured to put money into new receivers. Meanwhile, WQXR had to buy a new transmitter and start all over again to build an audience for FM.

Because WQXR had started as a pioneer in the art of realistic sound reproduction, it has always tried to keep abreast of progressive developments. The next step in perfecting the sound of music appeared to be something which, at the time, was called "binaural sound," which literally means hearing sound with both ears. When we look at an object, we see it in three dimensions because we get a slightly different image from each eye, and the brain blends

The WQXR-FM antenna atop the 54-story Chanin Building at 42nd Street and Lexington Avenue. In the background, the Empire State Building where WQXR-FM transmitter is now located.

them together to make a realistic picture. The same principle applies to our hearing mechanism. We hear a slightly different sound in our right ear than in our left, and the brain combines them to give a spatial dimension to the noise, speech or music we are listening to. That is why we distinguish directions from which various sounds are coming.

Before this, when one listened to a radio or phonograph, the sound originated from a single source, was picked up at one source and came funneled out of our radios from one point.

In the light of present developments in "stereo" broadcasting (about which more later), our experiments in binaural transmission seem rather crude. Our demonstrations were first made with "live" music. Two microphones were placed eight or ten feet apart in front of the musicians, each microphone representing one ear of the listener. The sound thus came from two different directions, and the vibrations picked up by the right microphone were fed into our AM transmitter and the left to the FM or vice versa.

At home the listener was instructed to use two sets and two speakers, one to receive the AM sound and the other FM. The listener sat between the two speakers and theoretically heard a three-dimensional performance. He certainly did if all elements were satisfactory—if his reception from our FM and AM stations was equal in volume and quality and his two receivers and speakers properly matched. These optimum conditions rarely existed in the average home, but the trials stirred up a great deal of interest. Again we were glad that WQXR was able to take the first steps in this development.

As interest grew in "binaural" reception, there was further proof that necessity is the mother of invention. It was known that in FM broadcasting it was possible to transmit simultaneously two or three different signals from the same transmitter by "piggy-backing" on the basic frequency. That became the principle of stereophonic sound or "stereo." Additions were made to our FM equipment which put us on the air as the first stereo station in New York. Meanwhile, the record manufacturers had developed stereo platters on which the same groove carried two channels of sound. Stereo pickups were developed which would register the differences in the channels and send them along to the FM stereo transmitter and finally to the stereo receiver in the home. The manufacture and sale of FM stereo home receivers and record players was a great boon to the radio and phonograph industries.

If any engineer reads this layman's explanation for layman, I am sure I will be taken to task for the oversimplification of a complicated scientific accomplishment. But this is a simple way to describe it, and I hope the technicians will forgive me.

Today stereo is a strong stimulus to FM broadcasting and a bonanza for the record manufacturer. Sound finally comes into the home the way we first dreamed about it when WQXR started as the "High Fidelity Station." Experienced people in radio at that time believed that people were not interested in the quality of sound but only in the content of the program. We believed that the two went hand in hand, and the success of stereo definitely proves that.

6

Records

THE CORNERSTONE of WQXR was good music, and good music to us meant records even though some people downgraded them as "canned music." We hoped and planned that in addition to records we would have actual performers, but we knew from the very start that we would stand or fall on the acceptance of our record programs.

Among the many discouraging comments we received in our formative period was, "Who wants to listen to records?" Indeed, that was the question record manufacturers were trying to answer for themselves in 1936.

Radio had almost eliminated the home phonograph with the result that the sale of records was at a very low point. People were spending so much time at their radios that they did not have much left for record playing. Also, the country was just pulling out of the Depression, and records cost money but radio, after you bought one, did not.

When WQXR was started, the total yearly sales of all record companies in this country was under $10 million. Ten years later, national sales were more than $70 million. Then the post-war boom in phonographs and records began, and by 1957, the figure was $336 million. In 1966, it was more than $700 million. Today it is a billion dollar industry.

The public has been purchasing records in amounts out of proportion to their other items of expense. In the decade ending with 1966, consumers increased their expenditures for all goods and services by 81 per cent, but they increased the amount of money they spent for recordings by 224 per cent.

Most of this great growth may be traced to the demand of teenagers and other young people for the records of their favorite personalities and groups. WQXR's repertory is largely in the fields of classical, light classical and a small proportion of original-cast Broadway show music. These three groups in the mid-1960's made up about 20 per cent in dollar sales of all records sold. Applying that percentage to total volume it would appear that the public was spending about $140 million annually for recordings of good music—more than 14 times the amount spent on all records in the 1930's.

In recent years there has been a decline in the sale of records of classical music. An article in *Time* Magazine for June 28, 1971, ascribes this falling off in part to the fact that the standard 18th and 19th century masterpieces have all been recorded dozens of times and most music lovers have them, and secondly to the phenomenal growth of "rock" music which has captivated most of the young generation. This same *Time* article, speaking of the young's taste in music says: "Given the right impetus, they are not necessarily averse to the classics—as proved by what *Elvira Madigan* did for Mozart's Piano Concerto K.467 or *2001: A Space Odyssey* did for Strauss' "Also Sprach Zarathustra."

Because of the success of WQXR in the New York metropolitan area, other good music stations started all over the United States and they, in turn, have influenced musical taste—although it must be pointed out that the definition of "good music" varies widely in different localities. Stations based on the WQXR image have been successfully built in Chicago, San Francisco, Los Angeles, Boston, Washington and in smaller cities, many of which are college or university centers. In almost every large community there is now a station which, though very broad in its definition of "good," is playing "middle of the road" music many cuts above the "rock" radio fare.

Hand in hand with the growing interest in good music was the technical improvement in phonographs, the long-playing record and the rapid acceptance of the stereo record which at last brought concert hall quality right into the home.

The broadcasting of records in the 1930's was under a cloud because every record bore a notice which stated that it was for home use only and not for broadcasting. There was legal opinion on both sides of this question, and it took many years to decide it in the courts. The record manufacturers were convinced that playing their records over the air would eliminate even their meager sales at that time; the performers felt equally strongly that no one would go to concerts or theaters or nightclubs to hear them in the flesh and thus the sale of their records on which they got a royalty would drop to nothing.

WQXR tried to convince the record companies, especially in the field of classical discs, that sampling their product on the air would create a demand. We pointed out that records were one of the few products you could actually sample by radio. They were obdurate and lawsuits continued to try to prevent the broadcasting of their discs.

Then some artists discovered that sales began to pick up in the vicinity of radio stations which were playing their recordings. Some began to write to WQXR that they would give us a release and save us harmless if we would air their records. This began to make an impression on the companies, and I recall that RCA Victor gave us a revocable license to play their records except for a few they might restrict from time to time.

Then came the historic Fred Waring case in which the court decided in favor of the broadcaster, and that settled the long controversy except for music which was restricted by copyright. The more records that were heard on the air, the faster the sales went and WQXR found that orchestras and artists urged us to play theirs. And what was even better for our growing station, the record manufacturers began to sponsor programs as a showcase for their new releases and started to pay for what in the past they had objected to receiving gratis.

The records were all at a speed of 78 revolutions per minute. Thus an album of a symphony consisted of many records, each side running only about four and one-half minutes. On a home phonograph, one had to get up and change the records frequently unless you owned one of those early record-changers which did not work too well and which had a bad habit of occasionally chewing up your favorite discs.

When people listened to an opera or a symphony played on WQXR, they were mystified because we gave them an uninterrupted performance for as much as 40 or 50 minutes at a time. They could not figure out how we could do it whereas they had the identical album and had to pause between every side.

Our method was a combination of equipment and skilled operation. In most cases WQXR had duplicate albums and used two turntables. The engineer who ran the turntables played one side of the record and, while that was on the air, he would "cue up" the first groove on the second side. As he saw the end of the first side approaching, he would start the second turntable but hold the record so that it would stay in the cued position with the pick-up already

An early control room with record turntable and console.

in the groove. Then as the last note on the first record sounded, the engineer would lift his hand from the second record and the composition would go on without a pause. This required much practice to develop a skill which usually made it impossible for one to notice the change from record to record. Occasionally, mistakes were made and by error sometimes a record might be played in the wrong order, which shocked the audience, but most of the time our engineers took pride in making the operation worthy of the great music we were playing. When the long-playing record became general, the need for such split-second timing vanished, but, to this day, people mention the mystery of the 78s.

Part of the Kardex catalogue of all the records in the WQXR library.

The attitude of the station with respect to records was an important part of our programming. The rule was that records must be presented with dignity. Every broadcast had to be prepared with an eye, or rather an ear, to how it would sound if it were in Carnegie Hall. If it was a symphony concert, it had to be balanced as carefully as a conductor would put together the numbers he was going to present at a concert. The rule was that there must always be unity in the broadcasts and some artistic reason for the choice. In the field of lighter music, the same care had to be taken to see that there was validity for the choice of selections.

To follow these policies, it was necessary to make an index of all records so that the programmers could assemble the proper music. It started with a Kardex file with a card for

A section of the WQXR record library.

each record. On that card was the name of the composer, the full title of the composition, the name of the individual performer or the name of the orchestra and conductor, and similar data on smaller groups as in the case of chamber music; the album or record number; its individual number in the WQXR library; and the exact timing in minutes and seconds. There was also a space on the card for entering the date of the last playing. The visible cards were filed in cabinets under the name of the composer so, for example, every composition of Brahms and information about it would be found in one place in the index.

This was the main source of information for our program department. In addition, there was another file system which grouped the records by the names of artists or performing groups. A third index listed records according to the playing time so that if we were looking for a piece of about 12 minutes, there was a place to find a composition of that approximate length which would be appropriate for the broadcast being planned.

Timing was a most important element because we had a firm rule that every program must fit into its scheduled time, and if a number had to be faded out because it was not correctly timed, the responsible person was made to realize that this was a serious error.

The care and planning involved a lot of work. This, in turn, meant a much larger staff than the average local station would employ. When we told people that our staff, not including news writers, was about 85 men and women, we were greeted by a surprised, "Why do you need all those people to play records?" As far back as the 1940's, the staff was made up as follows, not including management executives: announcers 10, engineers 17, time salesmen 6, staff musicians 10, continuity writers 6, music programming 6, clerical and miscellaneous 29. In addition, there were 11 who wrote and edited the hourly news bulletins, but that staff was on the payroll of the *Times*. . .

7

WQXR's Spot in the New York Picture

As PEOPLE became familiar with WQXR as a new kind of radio station, it was the "in" thing not only to listen to its programs, but to talk about them to friends to show that you were among the discriminating. There were all sorts of stories about us. Among the most amusing was one which appeared in the famous "Talk of the Town" section of the *New Yorker* Magazine, which was also in its salad days and of great influence among the sophisticated. The item read:

> A determined housewife of Park Avenue gave a dinner on the evening of Toscanini's final concert and after coffee and liqueurs led all her guests to the radio, which even as the party came into the room was issuing symphonic strains. All settled back in their chairs, and all had nobly restrained every little cough and rustle for quite a while, when they became aware that the lady's butler was signalling determinedly from the doorway. She held up a warning finger and pointed to the radio, but the butler didn't go away. He retreated about three feet outside the door, narrowing the field of vision, and resumed his gestures with insistence. Finally, the lady's husband, who is known in business and sports as a man of action, said, "Well, speak up, man. What is it?" "Beg pardon, sir," said the butler, "but you

> are listening to the wrong station. We have Toscanini in the kitchen and what you've got is WQXR.

The significance of this story, aside from the valuable publicity it gave us, is that it presented vividly one of the dilemmas which WQXR was trying to solve—whether our programming was going to attract a Park Avenue audience or a low-income group of musicians and music-lovers.

This was a serious problem for us because we had to know what our audience was like so that we could sell broadcast time to sponsors. In talking to advertisers and agencies in those early days, they agreed that we had a good station but said that it would only attract a small rich audience, and they couldn't sell chewing gum or corn flakes to them. They said that the station was great for Tiffany or Rolls-Royce. As an aside, it is worth noting that it took us about 20 years before we had Tiffany as a sponsor.

We were fairly well convinced that we had a large percentage of upper-income audience but certainly not one of millionaires. We wanted outside opinion, so we consulted a person who had a great reputation as a market analyst, particularly in the field of radio. After giving the problem some thought, he came up with the startling contrary conclusion that our listeners were "impoverished music-lovers" who he said, stood on lines for hours to get standing room at the opera and concerts and had very low purchasing power. He also delivered the discouraging opinion that not only was it a low-income group but also a small one, too small to warrant an advertiser to bother about using WQXR as an advertising medium.

These two expert points of view certainly put us between Scylla and Charybdis in trying to chart our course. So we spent money we could ill afford to find out through surveys and questionnaires what the facts really were.

In time we determined that, in a sense, both were correct. Just as in the *New Yorker* item, WQXR won its way into the kitchens as well as the salons. Over the years we have learned a lot about that audience which is certainly weighted on the side of better incomes. But we have also discovered that the common denominator is education and cultural background rather than wealth. Many of our listeners proved to have European backgrounds which accounted for their greater familiarity with serious music. In comparison with other stations, WQXR had far more intelligent listeners who were, in many cases, in the professions, business, educational fields, and generally leaders in their community. Their interests and habits filtered up and down to many other levels. That is an important reason why "word of mouth" advertising among our listeners built a substantial audience in size and purchasing power in a relatively short time.

The New York of the period of Fiorello LaGuardia may not have been as vast and frenetic as the city of John Lindsay, but it was the largest city in the United States and, consequently, not easily conquered by a new venture in its midst. WQXR entered upon the busy stage of New York with nothing but a new idea and with very little power or money to help sell it. We had to let the people know about us, and this in a metropolitan area of millions of families was not easy.

The newspapers did not pay any attention to this newcomer in radio, for they were devoting all the space they could afford to the big radio networks. To get a listing in the daily radio logs was difficult until a station had built a large audience, and to build this audience without such listings was almost impossible.

To break down this barrier, we decided to experiment with an advance monthly listing of the main programs which we would offer over the air and see if people were interested.

We got together a four-page folder which gave the featured musical composition of the broadcasts from 7 to 8 P.M. and 8 to 9 P.M. each day. The listing of the music on page 26 is the actual listing in the booklet for the week of June 1 to 7, 1936, the first week covered by the guide. We started to offer this pamphlet over the air in the middle of May at 10 cents a copy, and we received 63 orders the first day. This was encouraging because it showed that there were some people who would spend a dime and take the trouble to find out what we were going to play. When it became time to mail out this first issue, we had received 703 orders. Some listeners used the occasion of ordering the booklet to tell us how much they enjoyed the music. One of them sent a check for $15 saying, "You are more than deserving of the enclosed. Keep up the good work." This was the first of many checks that came as contributions through the years, and we always returned the money with thanks, but I could not swear that we returned this first one, for we were very short of money at the time.

For several months, until we could see whether the circulation would increase enough to make the service worthwhile, we kept asking people to send in 10 cents each month, but in a short time, we decided that our listeners wanted the booklet so we began to offer a subscription at $1 a year. At the end of the first year, we were mailing out more than 2,300 copies each month; at the end of five years almost 14,000 copies. The circulation continued to grow, and by 1947, it was about 47,000 and it continued to increase each year until it was more than 65,000 a month, which made it the largest monthly musical publication in the country. By this time, the Program Guide had become a little business of its own, but the subscription price of a dollar did not nearly cover our cost of printing, subscription clerks, mailing, postage and office space. We were forced to raise the rate to $1.50 a year, and this reduced the circulation materially. By then, WQXR schedules were listed in many newspapers and other publications, and the Guide was consequently not as essential to the listener.

W2XR

RADIO PROGRAMS

for

JUNE 1936

1550 K. C.
"At the End of the Dial"

STATION W2XR
41 PARK ROW
NEW YORK

The first issue of the program guide.

The Guide had for many years carried small advertisements, all of uniform size. These were not paid for, and each WQXR sponsor on the air who spent more than a certain minimum each week, received the ad gratis as an additional means of promoting his product or service. We considered opening the booklet to non-WQXR advertisers to meet the cost of production, but we decided that we were in the business of selling radio advertising on WQXR and did not want to compete with ourselves.

With the issue of December 1963, we discontinued publication, and as a substitute, the *Times* agreed to print programs of WQXR in the paper each morning.

In the course of its useful life, the little folder had grown to a 48-page booklet. In many homes, it was always near the radio, and many kept an extra copy in their cars so our programs could be followed while they were on the road.

In addition to being a convenient service, the program was of immense value to the station. First of all, it proved that our people responded to the right kind of advertising and to a product or service they wanted. It also enabled us to demonstrate the large size of the total audience. We engaged independent research companies to determine the ratio of subscribers to non-subscribers among our regular listeners. It showed that the ratio was 20 to 1 so that the circulation of the program multiplied by 20 gave advertisers strong evidence that we had a large family of regular listeners as well as many transients.

It also gave us something that no other New York station had—a list of names and addresses of listeners who were interested enough in WQXR to spend money to subscribe. When these addresses were spotted on maps, it showed that our largest circulation was in the better neighborhoods throughout the city and suburbs. It also demonstrated the power and coverage of WQXR, because it disclosed that we had a substantial number of subscribers in New England, Pennsylvania, Ohio, the District of Columbia, Virginia, North Carolina and Eastern Canada. These distant listeners could hear the station mostly at night when radio waves carry much farther.

The final issue of the program guide.

The subscription list was used many times to distribute questionnaires. Some of these were quite detailed, yet our audience willingly gave us information about themselves and their families. Most of the time, more than 90 per cent of the questionnaires were sent back promptly to the station for tabulation. In this way, we were always able to get a reliable sample of the musical tastes and characteristics of the WQXR family. We got to know who their favorite composers were, what kind of music they liked and at what time of day they liked to hear it. It also gave listeners a chance to criticize our programs, and that too was of great help to our Program Department. It was always our aim to broadcast the kind of music WQXR people wanted and not to try to force them to listen to things they did not enjoy. We told our staff to remember that little switch on the radio which is so handy and tempting to use. "There is no use broadcasting to silent radios," was our motto.

WQXR asked for volunteers among the audience to serve on an advisory committee. They were told in advance that they would receive frequent questionnaires, and the subjects would range from what they thought about Stravinsky to what kind of soap they bought; from chamber music to how much rent they paid or did they own a home; from their favorite operas to their travel plans. This group of about 3,000 is very conscientious and has been of real value to the musical and commercial planning of the station.

On the other hand, the advance information in the Program Guide helped thousands of people to plan their listening intelligently. It even helped with the planning of dinner parties, for I remember that people wrote us that they invited guests on certain evenings because they knew that the music scheduled for that time would please the company. Once in a while, when it became necessary because of something unforeseen to make last-minute changes, we would receive irate letters to the effect that Aunt Minnie came to the house particularly to hear the Mozart Jupiter Symphony and instead we played his Symphony No. 39.

One of the great intangibles of the monthly guide, particularly in the earlier years, was its creation of what we called the "WQXR Family." If you were a subscriber, it put you into an inner circle of music lovers. The appearance of the program guide next to your set showed your friends that you appreciated better radio. You had good taste. It also made you feel that you were a part owner of the station because of your dollar-a-year subscription. That was good for WQXR, for it made you a loyal listener and a missionary for our cause. As someone once said, "Everybody has two businesses—his own and radio." This was evident from the mail, particularly from some subscribers, who were sure they could run the station better than we did. Most of the time, the suggestions were impractical or useless, but every once in a while, we got an idea that we welcomed.

Most often they had schemes about how we could operate without advertising. Some said they would be willing to contribute $5, $10 or $15 a year to support the station. These well-meaning people had no idea of the cost of operating WQXR. Our entire organization had been unionized for all but the first few years we were in business, and that meant very adequate salaries for everyone from stenographers to musicologists. Even in the first 10 years, it cost more than $1,000 a day or more than $400,000 a year to cover our expenses. Some years WQXR's annual expense budget has been around $2 million.

We knew that the total number of generous people would not be enough to raise the amount required. The proof of this is that when we changed the subscription price of the program guide from $1 to $1.50 a year, the circulation fell off. How then could we expect enough people to pay much larger amounts to meet our costs? After the *Times* bought WQXR in 1944, there were some years when the newspaper had to make up substantial deficits and advance considerable amounts of capital for improving our facilities. There were other quite profitable periods for WQXR which enabled the station to pay back the money which the *Times* had advanced.

8

New Paths in Radio

There is so much talk these days about the universality of poor radio programs and what should be done to offer a greater variety of better broadcasts to the public, that it is interesting to look back and review what WQXR did in its own small way in the first few years of its life. Although we have emphasized thus far WQXR's dedication to better music on the air, the station felt the need to draw an audience from various other areas of interest and, to do this, we did some things which were considered daring then and which might even be regarded so today.

There was no "Fairness Doctrine" then, but WQXR acted as though there were. The large networks were quite cautious about what they broadcast because the late 1930's and up to the time that the United States entered World War II was a time of much bitter agitation. There were "America First" people at the right end of the political spectrum and the Communists at the other, and in between there were a lot of other "isms," all of whom wanted to preach their doctrines to the radio audience.

Looking through pre-war schedules discloses a great variety and contrast in our broadcasts. For instance, we broadcast a meeting of the "Committee for the Defense of Leon Trotsky;" another for the "American Friends of Spanish Democracy;" Norman Thomas differing with Associate Justice Black of the United States Supreme Court; a dramatic performance of "Waiting for Lefty" which resulted in complaints to the FCC; a series of broadcasts sponsored by the Socialist Party. In contrast during this same period were broadcasts of a Constitution Day meeting in Town Hall; eye-witness coverage of the American Legion Parade; luncheon meetings of the Foreign Policy Association; Bishop Donohue speaking about the election of the new Pope; political rallies of the Republican and Democratic parties; and a Yom Kippur service conducted by Rabbi Stephen Wise. There were many others also, and the only two applicants for time whom I recall we rejected were the notorious anti-Semitic Father Coughlin and Fritz Kuhn, the leader of the American Nazi Party. To many of the present generation, these names and movements may not seem significant, but in the late 1930's, they were dynamite. I will have more to say later about our pioneer broadcast on "family planning" and birth control by Mrs. Margaret Sanger.

This liberal approach gained many listeners for the young station and, at the same time, convinced others that we were the voice of Moscow. I recall one occasion when we had a WPA dramatic group on the air. This WPA (Works Progress Administration) group was made up of very competent actors and actresses who were out of jobs in those post-Depression years. The broadcast was a dramatized version of the "Song of Solomon." I received a telephone call at my home from a well-spoken lady who asked me what right did WQXR have to put Communist propaganda on the air. I replied that, to the best of my knowledge, the material was right out of the Bible. She agreed, but added that it was written by Jews, and all Jews were Communists. This more or less stunned me, and I said that if she knew the Jewish community in New York, she would realize that on the whole they were quite conservative and anti-Communist in their thinking. She came back with, "Don't tell me! I was married to a Jew for eight years!" That ended the conversation.

Our reputation for wide-ranging interests naturally drew a variety of personalities to the station to air their views and to present their ideas to an intelligent group. Looking through the records of those formative years, the names of people not in the field of music are numerous: James W. Gerard, a former ambassador to Germany; Alfred Landon, Republican candidate for President in 1936; Edward L. Bernays, the public relations expert; John Haynes Holmes, minister of the Community Church in New York and one of the leading liberal spokesmen of the time; Hendrik Willem Van Loon, author of many popular books on history; John Gassner, critic and lecturer on the theatre; Norman Thomas, head of the Socialist Party and many times Socialist candidate for President; Heywood Broun, columnist of the old *New York World,* famous liberal and founder of the Newspaper Guild; William Lyon Phelps, distinguished professor of literature at Yale; Theodore Roosevelt, Jr., and many more.

When the United States entered World War II, WQXR devoted a great many programs to further the war effort, to sell war bonds and to keep people informed by broadcasts which gave more background to the events of this crucial time than was possible in brief newscasts and commentaries. Many books were being published which supplied such in-depth background and Bennett Cerf, then a young publisher, volunteered to do a weekly 15-minute program called "Books Are Bullets" in which he interviewed well-known writers about their books and their ideas on the worldwide struggle. Some of the authors who appeared with Bennett Cerf (perhaps names which may not be as

Bennett Cerf and Bob Hope appearing on the "Books Are Bullets" wartime broadcast.

familiar today as they were in 1942) were Quentin Reynolds, Louis Fischer, Harold Denny, Walter Duranty, Louis Lochner, William L. White and Cecil Brown (all journalists), and authors Lin Yutang and Hendrik Willem van Loon.

This series was Bennett's first venture behind the microphone and it soon was apparent that he was a natural-born radio personality. The rest is history. He went on from his WQXR appearance to his many years on radio and television and became famous all over the country as one of the regulars on "What's My Line?"

Two other people who were with the station started their radio careers at WQXR and went on to fame in the field. One was Alistair Cooke whose pleasing British-accented speech was first heard over our facilities in a series of programs about the New York theatre and movies. His reviews were particularly interesting because he saw the theatre from the viewpoint of a British newspaper man and critic. He was a refreshing observer of what was going on on Broadway. When he left us, he became a popular master of ceremonies on the networks while continuing to broadcast regularly to Britain via the BBC—which he does to this day and is one of its favorite commentators because the British enjoy his incisive word-pictures of what is happening in the United States.

The other was a young publicity writer for a large motion picture company who relieved the monotony of his daily chores with an interest in poetry. His name was Norman Corwin, and he came to see me one day with an idea for a poetry program. Hardly a week went by when WQXR was not offered a poetry broadcast, but his approach was so novel and so far from the usual esoteric handling of poetry that we auditioned him on the spot, only to find out that his strong New Yorkese accent did not seem to go with the program he and we had in mind. I told him to go home and try to practice better diction and return when he felt he had improved. In about two weeks he was back, and he had conquered the defects.

We scheduled him to conduct a series which we christened "Poetic License," which indeed it was. It became a popular feature of the station overnight. The originality of his ideas and his production ingenuity in a few months resulted in CBS taking him away from us, for which I could not blame Corwin. On that network, he became the most famous radio writer and producer of his era. Those who were listeners in those days will surely recall the series "Ten by

Corwin," "On a Note of Triumph" and the moving tribute to President Roosevelt when he died. Corwin was undoubtedly the top man in serious, intelligent writing and production when radio was at the peak of its influence. When radio began to be put in the shade by TV, Norman Corwin transferred his talents to Hollywood. But he has never forgotten that WQXR gave him his start. When I retired from being Chairman of the Board of WQXR in 1967, Corwin wrote me a letter. It said in part:

> It is seldom that a man can put his finger on a moment of his life and say 'At that instant everything changed.' When our paths crossed thirty years ago, my life took another course. I choose to think it was for the better.
>
> God knows where I would be or what would have happened to me, had it not been for your generosity in giving what must have been the benefit of many doubts, to a brash young publicity flack from across town.
>
> . . . I think of myself as Sanger's Folly.

WQXR also experimented with several other unusual nonmusical programs. One of them was called, "Can It Be Done?" This was an idea brought in by a friend of mine, Alice Pentlarge, who said that there were many good ideas for new products which inventors tried to bring to the attention of manufacturers but had great difficulty in doing so. The plan of the program was to have one or more experienced business people listen to the would-be inventor explain his gadget. Frequently the expert had a chance to examine an actual model. There then was an on-the-air discussion of the practical value of the idea and its marketability. Some rather weird objects and odd inventors showed up, but some good ones also appeared and in this way found a way to the market. Meanwhile, the show held a novelty interest, and frequently there were laughs as people were reminded of Rube Goldberg's hilarious cartoon inven-

Another popular book program "The Author Meets the Critics." Left to right: Mrs. Eleanor Roosevelt, John K.M. McCaffrey, John Mason Brown and Ilka Chase.

tions. The idea was amusing but, like all novelties of this kind, it did not last long.

At the end of 1939, we met Dr. Henry Lee Smith who was an expert on speech and phonetics. Dr. Smith supervised a program which analyzed a person's speech so that Smith could tell where that person was born. We tried it out at an audition and then scheduled a regular series called "Where Are You From?" There was a studio audience for the show and volunteers were picked from the audience. Dr. Smith would ask the volunteer to say certain words. I remember that one phrase written out for the "victim" to read was "Mary–Merry–Marry". After these words were pronounced, Smith might say that the person came from the coastal region of Virginia, and he generally was correct. The next participant might read the same words, and Smith's diagnosis might be Vermont. He could even spot the different accents of the various sections of New York

City. He was so expert that he could sometimes tell whether you came from east or west of Central Park. This was a fascinating and amusing broadcast, each week bringing out new evidence of his skill. When war came the government tagged him to teach languages, and we heard that with his phonetic methods, he was teaching American boys to speak Japanese in a matter of weeks. The secret of the success of this program was undoubtedly its rare combination of an intellectual exercise and an amusing idea. The listener at home sooner or later found himself asking his friends, "Mary —Merry—Marry."

After the *Times* acquired WQXR, it was anxious that the station develop a program for young people which would be educational as well as a valuable public service. Some years before this, we had established a new program for children with Mrs. Dorothy Gordon. She had been featured on other children's programs on networks in the past. When we decided on a *Times* program for young people, we turned to her. We created the "Youth Forum" which gave young people, mostly of high school age, a place to discuss issues in the news and other subjects of particular interest to that age group. Dorothy Gordon was the leader of the discussions, and there was always an adult guest on the panel who was an authority on the subject to be discussed. The young people would ask him questions, and he, in turn, would have them express their opinions.

The "Youth Forum" which was generally broadcast from our auditorium on Saturday or Sunday mornings always had a studio audience of school children. This series was a fixture on WQXR for many years, and it was a public service of value to the New York school systems. It also served to familiarize young people with the *New York Times* and WQXR. Several years ago we discontinued it when Mrs. Gordon had a chance to take the program to NBC television.

Another non-musical program which attracted attention was a weekly 15-minute talk about books by Gilbert Highet, Scottish-born author and Anthon Professor of Latin at Columbia University. Professor Highet is a man of wide knowledge and great erudition, and his broadcasts, though sometimes intellectually pitched above the heads of some listeners, was a most stimulating experience. The series ran for several years in the early 1950's, first under the sponsorship of the Oxford University Press and later by the Book-of-the-Month Club. To give you an idea of the subjects he discussed, here are a few titles I have picked at random from a Program Guide of that period: "The Madness of Hamlet," "The Mysterious Dr. Johnson," "Books of Wisdom," and "Kitsch."

Gilbert Highet was not only a personality on the air for WQXR, but he was an enthusiastic listener. In a letter I received from him recently, he wrote:

> I remember how during my first few months in the United States in the winter of 1937-1938, I was suddenly taken ill and had to go to the hospital. Since I was prohibited from reading, my wife brought me a little radio. Accustomed to good music from the BBC in London, I was appalled by the Niagara of audible garbage which poured out of this instrument, and was becoming desperate when a fortunate turn of the knob took me right to the end of the dial, and suddenly a Mozart symphony filled the room. It came from WQXR, to which I have been a faithful listener ever since. Its music and cheerful voices of its announcers have helped my wife and me through many a minor trouble and one or two major ones. I am grateful to all the staff.

9

News and Public Affairs

We were sure that our audience wanted to be kept up-to-date on what was going on in the world around them. It was a time when crucial history was being made every day. Hitler was in power, and each year he dominated the scene more and more, culminating in the infamous Munich Pact in 1938. The networks were extending their news service to cover these great events, and we did not want our audience to forsake us for their broadcasts and forget to return to WQXR. Yet it was obviously impossible for a young and impecunious station to match elaborate network operation. Bear in mind that this was six years before we became part of the *Times*.

There was one way in which we might appeal to our hearers' taste for news. We decided that whatever small amount of news we might offer must be factual and not dramatized. The less reliable commentators and stations would frequently over-dramatize their delivery and facts and tie the news in with commercials. I remember many instances of this. One which comes to mind was in a story of goose-stepping German troops marching in review before Hitler. The commentator described the rhythm of their heavy boots pounding on the pavement and then said, "and speaking of shoes, there is a great sale at———shoe stores this week."

Having been trained as a newspaper man, I was disturbed by our inability to make news an important part of our schedule in those early days. The networks had established their own staffs of reporters and commentators because, at the time, newspapers were unwilling for the Associated Press, which they controlled, to supply news to radio which the papers regarded as a strong competitor. It took the publishers (including the *New York Times*) years to realize that hearing brief news items on the air would stimulate more people to buy newspapers for greater detail and authoritative background. As a sop, the newspapers cooperated in the creation of the Press Radio Bureau which was to issue condensed summaries a few times each day. I do not remember how much we paid to get five minutes of this each day (none on Sunday), but I do remember that we could not afford a ticker so we had to send a messenger for it each afternoon. It went on the air from 6:45 to 6:50 P.M., and later in the year when we had some morning broadcasts, we had another five-minute summary at 10:45 A.M. The first major use we made of Press-Radio was for the results of the 1936 national election for which event we installed a teletype so we could give returns on the Presidential and other contests interspersed with music throughout the evening. This format, combining news with music, evidently pleased the WQXR audience, for at midnight we asked over the air whether we should continue the service. Immediately the telephone switchboard was overwhelmed with calls, so we continued into the early morning.

In order to add more news without added expense, we arranged with the *Christian Science Monitor* to broadcast its daily radio script of news background which was well written and came from that highly respected newspaper. Best of all, we did not have to pay for it, and even the announcer was supplied free by the *Monitor.* The announcer was Rex Benware, whom we later hired as a full-time member of the staff and whose voice was familiar to our audience for many years thereafter.

Another improvisation was a nightly 15-minute talk by someone in the public eye at the moment. It originated from the Hotel Roosevelt in New York. The hotel supplied the room which we set up as a remote studio, in exchange for the mention that the program came from the Roosevelt. Such were the simple economics of the day. Many speakers were glad of an opportunity to air their opinions, and we were careful to give differing points of view a chance to be heard. I remember that the first personality on this program was Representative Hamilton Fish, a conservative Republican, very much opposed to F.D.R. and the New Deal. His subject was "Social Security" which was just getting started, and you may be sure it was not to Mr. Fish's liking. One of the important by-products of this series was a daily listing of these newsworthy broadcasts in the Radio Program Highlights of the *New York Times.* WQXR needed promotion of this kind, and it helped swell the ranks of our listeners.

By the last month of 1936, we had added a daily news commentator. He was Percy Winner, an experienced journalist and foreign correspondent whose familiarity with the European scene was important in analyzing the news as Hitler grew in power and the outbreak of war in 1939 grew nearer.

The next year we presented another commentator, Robert Emmett MacAlarney, for many years a newspaper man in New York and city editor at one time of the *Tribune.* He had been my favorite professor at the School of Journalism at Columbia, and I had kept in touch with him since my graduation. In addition to his background and wide experience as a newsman, he had a most winning on-the-air personality, and he was an immediate success with our listeners. He undertook this daily job without much hope of financial reward, but he did it because he believed in WQXR and what it stood for. It was a great personal satisfaction to me to be associated again with my former teacher.

We wanted to keep our audience informed about what was going on in their city as well as with world news, and one of the ways of doing this was to broadcast public dinners and other events. This was an inexpensive feature because we had plenty of unsold time and the sponsoring organization for the dinner or meeting paid WQXR for the rental cost of the necessary telephone lines. One of these early occasions was a talk by Mrs. Margaret Sanger (no relation of mine) from Town Hall. Her subject was birth control, she being the great pioneer in the movement. For WQXR to run the risk of being violently criticized for permitting this subject to be discussed on the air was one we took with our eyes open. Actually, there was little criticism of us, and I look back with satisfaction to our initiative in being the first station, to the best of my knowledge, to ever broadcast a talk on the need for family planning and birth control.

The years immediately before the start of the war in Europe were marked by a demand from intelligent people for thoughtful analysis of the worsening world situation. WQXR saw a need to supply it, and we were fortunate in being able to present, during these tense years before and after Hitler's invasion of Poland, two analysts who fit the bill. They were Quincy Howe and Lisa Sergio.

Quincy Howe was and still is a well-known writer and broadcaster on foreign and domestic affairs with many years experience as an editor. He was a member of the distinguished Howe family of Boston, and his voice betrayed his New England heritage. He had an intimate manner of delivery which made the listener feel that Howe was talking to him and not pontificating from on high. His approach to the startling news of the day was not sensational, and although his point of view was more liberal than most network commentators, no one could accuse him of being radical. Howe's audience liked him, and it grew so fast that after about two years on WQXR the Columbia Broadcasting System made him an offer to come there. He hesitated because he liked working in the atmosphere of WQXR, but the offer that CBS made was one we could not match, and he became a network commentator. WQXR's listeners missed him and Howe missed WQXR, for he found that he did not have the freedom at CBS that he had had with us. After a while, he decided he was not comfortable on the network and he departed. During his time at our station, Quincy and I became good friends and that has lasted to this day. We often lunch together to talk over "the good old days."

Lisa Sergio was a most attractive personality both on and off the air. Italian-born, European-educated, she had a personality that was different from other commentators and a European background, part of it in Fascist Italy, which was essential at this time in interpreting what was going on between Hitler and Mussolini. She had known Mussolini well and had found it advisable to leave Italy when she violently opposed what Il Duce was doing.

Lisa Sergio had a great command of English and the perfect diction which so often puts to shame the native American. Her comments on the current scene, couched in her beautiful delivery, soon attracted a large audience. Not only was she worth listening to on foreign news developments, but her background in music, art and other cultural things always added interest to her broadcasts.

Having a woman news analyst was somewhat unusual but it was not unheard of, for Dorothy Thompson, at one time the wife of Sinclair Lewis, was one of the most popular and authoritative columnists and radio personalities. In fact, some people thought that she and Lisa Sergio were one and the same because their air personalities were somewhat alike. Lisa was an Italian and a woman, a combination which occasionally resulted in a show of temperament. She and I had several scenes where we did not see eye to eye, and that often led to an explosion on her part and sometimes on mine. But we always ended up with a cooling-off period, and she was a valuable asset of WQXR for some years.

August 1939 was a month of climactic crises in Europe, and the need for fast and accurate news reporting became greater than ever. For some time we had been discussing with the *New York Herald-Tribune* a news arrangement between it and WQXR. It came to fruition on the first day of that fateful month, and beginning at 11 P.M., we broadcast a summary of the day's news direct from the *Tribune* offices on West 40th Street. A member of the *Tribune* staff broadcast the report every night at that hour. This was an arrangement which was useful to both the newspaper and the station because each gave the other helpful promotion, and it also gave our audience an exclusive source of news from a reliable newspaper at a time when people were hungry for it.

In our preliminary discussions with the *Herald-Tribune*, we had talked about the newspaper acquiring a financial interest in WQXR as a logical alliance between a quality newspaper and a quality broadcaster. The nightly news broadcasts were an exploratory step. Because New York newspapers were still suspicious of radio as a strong competitor for advertising dollars, the management of the *Tribune* decided against any form of merger, and the whole project, as well as the news broadcasts, was dropped after a few months. In view of the purchase by the *Times* of the station less than five years later, it is interesting to speculate on what might have happened if the *Tribune* had acted favorably.

During the period just before and after the United States entered the war, there was evidence of interest in WQXR from several publishers. The *New York Post* talked to us about buying all the stock of Interstate Broadcasting Company, but Jack Hogan and I, though willing to sell them a minority interest, were not going to sell all. Another newspaper, New York *Daily News*, had several talks with us, but nothing developed.

But in 1942, Time, Inc. decided that radio was a field they should explore. They had at that time no experience in communications other than their very successful magazines. As WQXR wanted to expand its service in news broadcasting, an alliance with *Time* seemed logical. We agreed on mutual cooperation, with *Time* preparing and presenting certain programs subject to our acceptance; we to supply the facilities. In order to strengthen the operation, *Time* invested $100,000 and received preferred stock in WQXR which could be converted by *Time* into common stock upon a further investment at a future date. Our first joint venture was a news program of 15 minutes five days a week at 9 P.M. The news report, written in typical *Time* Magazine style was excellent. It was delivered each evening by Westbrook Van Vorhees whose voice was famous for his "March of Time" appearances in the movies and on radio. Later *Time* and WQXR experimented with a new idea called, "Let's Learn Spanish". This was a series of lessons which combined the use of a book which the listener followed as he listened to the broadcast. It was well-received, and the series was syndicated by *Time* to other stations outside the range of WQXR. Two members of the *Time* organization became directors of Interstate and were helpful to the station in many ways. This joint effort was a pleasant association and continued until the *New York Times* bought the station.

10

The Sale to the *Times*

When we were getting organized and looking for a location for the WQXR studios, Jack Hogan and I went to see Arthur Hays Sulzberger, president and publisher of the *New York Times,* to ask if he was interested in having the station in the Times Building. He was his usual affable self and listened to our story. He said that as soon as we moved into their building, even as a tenant, people would think that the *Times* owned us, and he was not inclined to get the paper involved with a radio station. This was in 1936.

A few years later when the station was very much in need of financial assistance, we again approached him and offered to sell the *Times* a minority interest for less than $50,000. By that time he was more familiar with the station, and whereas he thought we were doing a much-needed job in radio, he still felt that the *Times* did not want to go into radio.

When the *Times* would not take us in as a tenant, we hunted for studio space in the mid-town area of Manhattan and decided on the Heckscher Building at 57th Street and Fifth Avenue where, over the years, we added to our faciliities and where we remained until 1950 when we finally moved into the Times Building.

So there was no further contact with the *Times* for some years, and in that period, the station became successful and a more important part of the New York scene. During these years WQXR had been able to expand its transmitter power and its programming by taking in more stockholders. By the issuance of several kinds of stock and options, we had raised sufficient capital to reach a point by 1942 where we were making a moderate profit.

Much of the refinancing was done under the helpful guidance of Ralph Colin who had been my personal attorney even before I was in the radio business. Ralph Colin was more than a legal and financial advisor because he knew a great deal about broadcasting. He was a director and the general counsel of the Columbia Broadcasting System from its earliest days.

The added stockholders were, for the most part, family and friends of Jack Hogan and mine, and many made investments in our highly speculative venture because they had confidence in us. There were fewer than 40 stockholders, and Jack and I still held 75 per cent of the voting stock of the corporation. For that reason, the Federal Communications Commission did not have to concern itself with the ownership and control of the station.

In January 1944, one of our directors, a well-known attorney and artist, James N. Rosenberg, told Jack Hogan and me that he had an offer of $1 million for the station, but it would have to be for 100 per cent of the stock. This was the highest price ever paid for a radio station up to that time, and Jack and I could not dismiss it. We both felt that it was fair to both of us, and it also would give us an opportunity to insure a good profit to all those who had shown confidence in us by their investment.

The next day we met with Morris Ernst, a distinguished attorney always active in liberal movements. He said that he was speaking on behalf of his client, Mrs. Dorothy Thackrey, publisher of the *New York Post*. He outlined the proposition in greater detail and added that the plan included an attractive salary for Jack and me, each of us getting five-year contracts of employment.

In the next few days there were other meetings and at one, my wife Eleanor was present as program director. We talked over general policies with Mr. and Mrs. Thackrey. Everything was most amicable and Mrs. Thackrey (now Mrs. Dorothy Schiff) showed great interest in what we were doing but did not say much about her plans for the station, although she inferred that we would be able to continue our established policies.

My diary indicates that this meeting was on a Saturday, and the next day Jack Hogan and Ralph Colin came to my home, and they with Eleanor and me considered the pros and cons of the important step we were about to take. Although Jack was very much in favor of the deal, I frankly said that I had misgivings. I said that the *Post* was not of the same quality as WQXR and that I was sure that if the *Post* owned the station the image of WQXR would change for the worse sooner or later. I also said that whereas I liked the Thackreys, I was worried about what they might do to the station. At the same time, I emphasized that as a minority stockholder, I could not block the sale, nor did I have any intention of depriving all our faithful stockholders of the profit to which they were certainly entitled in view of the risks they had taken. So I said, "I'll go along, but I wish it were the *New York Times!*" . .

Those were the fateful words. The next morning Ralph Colin came to my office and asked me whether I really meant what I said about the *Times.* I replied that I certainly did, whereupon he went to the telephone and called Louis Loeb, general counsel of the *New York Times.* He told him briefly what we were thinking about, and they agreed to meet the next morning and ride down in a taxi to their offices and discuss it en route.

Ralph and I then met with Jack Hogan, and both Jack and I insisted that this negotiation should not become an auction between the *Times* and the *Post,* but the offer would be made to the *Times* on exactly the same terms. We did not want a penny more nor would we take a penny less. With that injunction, Ralph said he would talk to Louis Loeb.

As a result of this talk, Louis Loeb spoke to Arthur Sulzberger and his son-in-law and assistant, Orvil Dryfoos. They were interested and knew that they would have to make a quick decision. Arthur Sulzberger generally conferred on important matters with Gen. Julius Ochs Adler, then the Vice-President and General Manager of the *Times,* but the General was with his Army division in Southeast Asia where the fighting was going on. In the years since WQXR had broached the subject of radio to the *Times,* there had been a change of attitude. The *Times* by now realized that it would be advantageous to have a news service on the air to promote circulation and as a good-will service to the public. A few years earlier, Arthur Sulzberger had made an arrangement with his old friend, Nathan Strauss, owner of station WMCA, to broadcast *Times* news bulletins every hour on the hour over that station, and fortuitously the *Times* started the broadcasts on December 1, 1941, less than a week before Pearl Harbor.

In that way, the *Times* had broken the ice in getting into radio and had found out by 1944 that its radio news service was helpful to the paper. But Sulzberger and others at the *Times* were not pleased with the general atmosphere of WMCA whose programs certainly were not in the spirit of the *Times.* Sulzberger knew that the *Times* would feel more at home at a station which more nearly reflected the image of the paper, and WQXR was that station. Thus it was easier for him to make up his mind quickly, and within 24 hours, he said the *Times* would buy WQXR on the identical terms offered us by the *Post.* Fortunately, we had made no firm commitment to the *Post,* and we told Morris Ernst that the deal was off. He was very upset and annoyed at Jack and me, for he was most anxious to acquire the station for his client.

At a meeting on January 25 with the top officials of the *Times,* everything was agreed to, but we did not want the news to become public until the arrangements had been reduced to writing. But the next day, Leonard Lyons in his column in the *Post* wrote: "Radio Station WQXR was sold to the *New York Times* yesterday, subject to the approval of the Federal Communications Commission. The price was over a million."

That brief item created lots of excitement, especially among the staff of WQXR, and Jack Hogan and I issued a memorandum saying that talks were going on between the *Times* and us, but there was at yet no final agreement. We assured the staff that if the transaction were consummated, it would not change the nature of the station nor the status of the management and staff. With this assurance, things quieted down and the lawyers and principals worked late at night and over the weekend to complete the formal agreement. By Tuesday, February 1, 1944, everything was in order and at 6 P.M., the agreements were signed and the story released for the next morning's papers. The announcement said in part:

> The New York Times has today agreed to purchase from Mr. Hogan, Mr. Sanger and their associates all of the stock of Interstate Broadcasting Company which operates Station WQXR and Frequency Modulation Station WQXQ. The purchase is subject to approval by the Federal Communications Commission.
>
> Mr. Sulzberger stated that Mr. Hogan and Mr. Sanger will continue under five-year contracts as the chief executives of the broadcasting company and he stressed the fact that the Times

> did not contemplate any change in the station's personnel or program policy. 'The Times is proud to acquire a broadcasting station which throughout its history has constantly maintained policies emphasizing programs of high standard and unique quality,' he said.
>
> Messrs. Hogan and Sanger stated that they would not have been willing to sell except to a purchaser which, like the Times, was determined to continue the essential character of the station.
>
> The Times' news bulletins "Every Hour on the Hour" now broadcast over station WMCA will be continued over that station for the time being.
>
> Mr. Sulzberger also announced that Mr. Nicholas Roosevelt will be the liaison executive between the Times and its broadcasting interests.

Nicholas Roosevelt had been with the *Times* and left more than 13 years before. In the interim, he had served as United States Minister to Hungary, an editorial writer on the *Herald-Tribune* and, during the war, he had the important post of Deputy Director of the Office of War Information under Elmer Davis. His selection for the liaison job was most appropriate. As a young man, he had been close to his much older cousin, President Theodore Roosevelt and, in that way, had made political, social and cultural contacts in many parts of the globe. As a lover of good music he was enthusiastic about the standards and purposes of WQXR, and his advice and assistance to us was most helpful. He was only with WQXR a few years because he wanted to retire to his home in California to write, but in the short time of our association, my wife and I became close friends of Nick and his wife, Tirzah.

On the day that the proposed sale was announced, Arthur Sulzberger sent a letter to the staff of WQXR. In part it read:

> When the new ownership of the station is confirmed, there will be no changes in program policy, and no changes in staff or operations are in view.

> We believe that we know how to publish a newspaper, and we respect the skill that you have displayed in building a radio station that has won the admiration of its listening audience.

These two sentences are significant, for the attitude expressed in them was the basis of the operation and relationship of the newspaper and the station for the more than 20 years that I was at the helm.

One interesting sidelight of the sale. The agreement with the *Times* provided that we were required to deliver every share of Interstate's stock because the *Times* did not want any minority stockholders. There was not much trouble in getting the common stockholders to agree to the sale, for they were making a very good profit on their investment. The only stockholder that would not make a profit was Time, Inc. which owned $100,000 of preferred stock for which they would receive their investment in full but no more.

I went to see David Brumbaugh, the Treasurer of Time, Inc. and told him of the proposed deal and that all the stockholders except *Time* would make a profit but that *Time*'s preferred stock would simply be redeemed at face value. I also told him that the *New York Times* insisted we deliver every share and that he was in a position to block the sale but that I hoped he would not. He thought a few moments and then said, "Elliott, do you want this deal to go through?" I said that I certainly did. "Okay," he said, "we'll go along."

11

"The Radio Station of *The New York Times*"

The agreement with the *Times* was signed on February 1, 1944, but the sale had to be approved by the Federal Communications Commission. This was a time when there was much agitation about newspapers controlling radio stations—today still an unsettled controversy. There had been an increasing trend toward newspaper ownership of stations within their own circulation areas. In some towns where there was a single newspaper, the lone radio station was under the same ownership. Many people felt that this trend would lead to a monopoly in the distribution of news and opinion and thus restrict freedom of expression in such communities. Liberal organizations were trying to get Congress to prohibit newspaper-radio alliances, but Congress had taken no action, possibly due in part to the fact that many members of the House and Senate had investments in newspapers, or radio, or both.

In the first few days of the negotiations with the *Times*, Jack Hogan telephoned to James Lawrence Fly, then chairman of the FCC and told him of our intentions. Fly's reaction was that in view of the standing of WQXR and the *Times*, he saw no objection on the grounds of newspaper ownership, but he added that, of course, he could not speak for the other members of the Commission.

It took about four weeks to prepare the application for the transfer of our license which involved documents and reports about WQXR as well as everything about the ownership of the *Times,* even including the will of the late Adolph Ochs, founder of the present *New York Times.* After sending this voluminous data to the FCC, we had to wait patiently for a decision.

Months went by, and we began to wonder what was going to happen. At this critical period, the *Times* published a column by Arthur Krock, its chief Washington correspondent at the time, in which he took the FCC to task for some of its recent actions. Although the *Times* realized that the publication of this criticism might result in an unfavorable decision by the Commission, it did not hesitate to print it.

Finally on July 18, 1944, the FCC approved the sale with only one of its seven members dissenting.

WQXR had a young man working for us as a junior salesman who had been in the classified advertising department of the *Times* but had resigned from that job several months before. He had been hired by the station before there were any negotiations with the paper. When the deal was completed, his jocular comment was, "Gosh, the *Times* went to an awful lot of trouble to get me back!" That young fellow's name was Robert Krieger, who later became the station's vice-president in charge of advertising sales.

But not everybody was so enthusiastic. Some of the people at the *Times* still were not sure that the paper should be in the radio business. Some thought broadcasting was not quite respectable. Some of this antipathy stemmed from their opinion that radio was an interloper, particularly in the field of news dissemination. Then there was another group who were not opposed to radio, but their idea of a radio station was not WQXR. They did not think there were enough people who wanted good music and superior programming. There were also those who just did not like good music.

Fortunately, Arthur Sulzberger and Orvil Dryfoos believed in our policies, and they always backed the station to the limit and that is what counted. At the time of the purchase, the then Promotion Director of the *Times,* Ivan Veit, was an officer in the Navy serving in the Pacific. When he came back after the war, he was enthusiastic about the acquisition. He saw in WQXR an ideal means of circulation promotion for the *Times* as well as an extra service to *Times* readers which no other newspaper in New York was supplying. In addition, he was a lover of fine music so he had no hesitancy in urging us to more and better music rather than encouraging us to lower our standards as some would have had us do. His support of the station and his guidance through the years was an important part of our growth and now in addition to being an Executive-Vice-President of the *Times,* he is also President of WQXR.

As is well-known, the *Times* is conservative about making changes. It was even more cautious back in the 1940's. The management of the paper made it clear to us and to the public that the *Times* wanted to feel its way into radio, and so no marked changes would be made in its relationship to WQXR for some time to come. This made it difficult for me who saw the worldwide facilities of the newspaper lying there, and WQXR not being able to adapt them to radio by the use of *Times* personnel on the air.

It was particularly galling to hear the *New York Times* news every hour on the hour on a competing radio station. But the *Times* had a contract with WMCA for the exclusive use of *Times* news on the air, so WQXR had to content itself with the regular run of Associated Press service even though it was "The Radio Station of the *New York Times.*"

This situation continued for about two years, and after a while, the *Times* became just as impatient as we were. As WMCA developed its programs into the mass music field to reach a larger audience, the *Times* felt more and more like a fish out of water. There was some acrimonious correspondence about this between Arthur Sulzberger and Nathan Straus, the owner of WMCA, and it became obvious that a change had to be made. It was eventually agreed to part company as of July 1, 1946. The change was announced in May, and with it, the *Times* stated that the news on WQXR would not be offered for sale to any advertiser but would be sponsored by the *Times* itself. The *Herald-Tribune* took over the news service on WMCA as soon as the *Times* moved to its own station.

It was typical of the *Times* to take a stand against the sponsorship of its news programs because there was a feeling that if news were sponsored, the listener might imagine that the news content was influenced by the advertiser. This same policy continued for many years at a great loss of revenue for the station and at great expense to the *Times*. As the *Times* news on the air came to be regarded as one of the best radio news services, there were numerous offers by advertisers to sponsor it.

My WQXR associates and I were strongly of the opinion that there would be no breach of journalistic ethics if the *Times* had complete control of the news, and no one on the station or with the advertiser or advertising agency had any influence on the radio news department. We felt sure that no one would have the effrontery to try to tell the *Times* what to put in or leave out of the news broadcasts. I argued with Arthur Sulzberger that the separation of news and advertising could be maintained by keeping the function of the station and the function of the news department entirely apart. I suggested to Orvil Dryfoos that each broadcast should be introduced by a sentence such as "Here is the news, prepared and edited by the *New York Times* and brought to you by the XYZ Company." That seemed to be the formula that satisfied everyone. The station received the green light to go ahead, and the sponsorship of the *Times* news broadcasts has become an important source of revenue to WQXR.

In the thousands of sponsored newscasts, I recall no instance where an advertiser has even so much as hinted that an item should be included or excluded, nor do I recall any complaint from a sponsor that a story affecting them was inaccurate or unfair. There have been instances where the *Times* news has included items about labor relations or other matters which were not favorable to the advertiser though entirely truthful and accurate. Sponsors understood that *Times* news and WQXR advertising are two separate worlds.

Once the *Times* news was on the station, there was a gradual relaxation of the policy against the use of *Times* personnel on the air. The first important development was elaborate radio coverage of election returns for which the *Times* was so well equipped. The election of 1948 was our first chance to do a radio reportorial job worthy of the *Times*. Bear in mind that 1948 was the last Presidential contest before television more or less supplanted radio as the most popular source of election returns. The 1948 election was the Truman-Dewey battle which people knew would be close, though probably ending up with the election of Mr. Dewey.

We had a team of *Times* and WQXR men on the air headed by the late William Lawrence (subsequently an important news commentator of the ABC network) for the *Times*, and Bill Strauss, veteran announcer of WQXR who had been the station's election night voice for many years. The early returns were coming in as expected, and it looked like a victory for Dewey. The *Times* was preparing to announce his victory in the first edition when James Hagerty, Sr., for many years the *Times* expert election prognosticator, advised the editors to hold up any concession because he had noted certain trends in the voting in upper New York State which if continued would indicate that Dewey did not "have it in the bag." So the first editions said the election was close, and so did WQXR on the air. You will recall that this was the election night when the *Chicago Tribune* came out in its first edition with a big headline which read: "Dewey Elected." President Truman delighted in showing this front page after the result was known.

WQXR remained on the air all night as the seesaw battle went on, and people stayed at their radios because they knew they could rely on the *Times* reports via its own station. After working all night, our exhausted staff had the satisfaction of still being at their microphones and typewriters when Dewey conceded defeat at 11 A.M. the next day.

The cooperation of the newspaper and the station in the coverage of this important election did much, I believe, to convince the few remaining skeptics at the *Times* that the paper really had a reason to own the station.

There were many other occasions when the station became an adjunct to the newspaper in distributing the news. For example, on July 26, 1956 I was awakened at 2:40 A.M. by Orvil Dryfoos who telephoned to tell me that the liners "Andrea Doria" and the "Stockholm" had collided off Nantucket. The station had signed off for the night at 1 A.M., and Orvil asked me to get the station back on the air as fast as possible to carry a running account of the disaster. I was at my summer home in Westchester County, and I got on the telephone and woke up the heads of the various departments of WQXR and told them to get their staffs together and get on the air as soon as they could. In an hour and 20 minutes, a skeleton crew was assembled, and we started our broadcasts at 4 o'clock and gave continuous news coverage so that *Times* readers and others would be kept informed even after the last edition of the *Times* had been printed and distributed.

A similar case of emergency coverage took place when the news broke about the Soviet intervention in Hungary to put down the ill-fated revolt in that country. We stayed on the air all night, for at the moment the story broke, it was conceivable that war between the United States and the Soviet Union might break out any minute. When the U.N. Security Council went into session at 3 A.M., we broadcast the discussions direct from the meeting.

WQXR's most important emergency service to the *Times* and to the public came in the long strike periods when the paper was not published. There was a 19-day strike in December 1958 which turned the *Times* into a newspaper of the air via WQXR. In that period, the station aired 4,920 minutes of news and features in 10 and 15 minute segments every hour on the hour and at a number of half-hour intervals. Our regular musical programs, somewhat abbreviated, were scheduled between the news coverage. An augmented broadcast department in the *Times* news room wrote and coordinated all broadcast material under the supervision of Clifton Daniel, later the managing editor of the paper. Correspondents from all over the United States and abroad telephoned in their stories which were taped and put on the air. There was a direct line to the *Times'* Washington bureau for a prompt flow of news from the capital.

Of our coverage, *Time* Magazine said, "Of all the radio and TV stations that tried to fill the news gap by extended coverage, the best job was done by a radio station tied to a good newspaper—The *New York Times'* WQXR. *Variety,* the show business weekly, said, "True, there were a handful of redoubtable broadcast performances turned in during the strike. As might be expected under such circumstances, they stemmed largely from a newspaper-owned station—WQXR, The *N.Y. Times* radio sentinel. With savvy showmanship and scholarship, the *Times* news and editorial staffers performed absorbingly and authoritatively. By comparison, the other stations were struggling with a pallid roadshow version."

When the 114-day strike took place in 1963, the pattern of news coverage in substitution for the paralyzed *Times*

THE 1950 GEORGE FOSTER PEABODY RADIO AWARD

FOR OUTSTANDING ENTERTAINMENT IN MUSIC HAS BEEN WON BY WQXR

"...no station anywhere has devoted more time or more intelligent presentation to good music than has WQXR"

THE JUDGES, IN MAKING THIS HIGHEST AWARD IN RADIO, CITED WQXR IN THESE WORDS: "Radio generally has done much to increase and uplift musical appreciation in this country. But no station anywhere has devoted more time or more intelligent presentation to *good* music than has WQXR. All types of the best in music—instrumental, chamber, solo, opera and symphonic—have been brought to half a million families in New York alone, plus homes in fourteen other states and Canada. And the performing artists have been a veritable 'Who's Who' in the world of music. Prominent in 1949's offerings was the 'Our Musical Heritage' Series. In recognition not merely of this and other programs, but primarily to single out and honor the station for its overall contribution to musical appreciation and good music, the George Foster Peabody Award for outstanding entertainment in music goes to WQXR of New York City."

In addition to winning the Peabody Award for news in 1963, WQXR had also won it for music in 1950.

was restored and improved by our experience of a few years earlier. This time WQXR received the top prestige honor of broadcasting, the Peabody Award. The citation reads, "Consistently excellent in its news coverage at all times, WQXR, New York, merits special praise for lighting a candle in the darkness every night during the New York newspaper strike with its concise, authoritative digest of the day's news." The citation and medal were presented at the Peabody ceremonial by my old friend Bennett Cerf, to my son, Elliott, Jr., who had been responsible for coordinating the operations of the paper and the station during the emergency.

Finally in January, 1971 the *Times* gave its blessing to the use of its correspondents on a regular basis for a half hour daily news program called "Insight" with Clifton Daniel, Associate Editor of the *Times,* as "anchorman." Until Daniel left New York to become head of the *Times* Washington bureau, the program was heard Monday through Friday evenings and my son, Elliott, Jr., was its executive producer.

"Insight" was enthusiastically received by a large audience and undoubtedly was the most successful news program WQXR had ever developed. Unfortunately, no one else with Daniel's news and broadcast experience was available at the *Times* so "Insight" was dropped after the Presidential election of 1972.

In each program Daniel talked on the telephone or interviewed in the studio, members of the staff, one at a time, discussing the important news of that day. Daniel on the phone to Tokyo, Washington, Hong Kong or wherever big news was breaking, talked to the correspondent, asked questions and brought out the background and significance of the event so that listeners gained a real understanding of what had happened, what might happen and what it all signified. These conversations were entirely *ad lib* and were most effective because they were informal talk between one experienced newspaper man or woman and another.

12

"America's No. 1 Fine Music Station"

I MUST admit that we took some liberties in adopting that slogan. But it was not done in a spirit of boastfulness. Many people described WQXR that way, and it appeared in print so often that we felt justified in using it.

In the New York area, we had little good music competition. New York City's municipal station, WNYC, programmed some classical recorded music even in the earliest days of WQXR, but being the city station hindered it from competing successfully with us. WNYC had the civic obligation to transmit some rather humdrum material in doing what the city thought was the station's function to let the people know the price of eggs; what the Department of Sanitation was doing to clean the city's streets; who was being honored with the keys to the city at City Hall; and many so-called educational programs. Also WNYC was forced to rely somewhat on volunteer and sometimes amateur talent because the city fathers were not generous with the station's budget. For all those reasons, people could not rely upon a uniformly good flow of music without interruption for more mundane material. On the other hand, WNYC had one great advantage over WQXR. The Musicians

Union, for policy and political reasons, for years permitted WNYC to broadcast concerts and other musical events which were denied to us because of exorbitant union scale talent fees. The city station paid nothing for these even though it then employed no staff of union musicians.

This situation seemed unjust to us, and though we protested to the union, we made no progress. It was the old adage, "You can't fight City Hall!".

The existence of WNYC was good for us because it supplied a yardstick to measure how well or poorly we were doing in the community. In the early years of our existence, the WNYC audience had to learn that WQXR was on the map, but once they became familiar with our programs, they preferred our more professional radio production and content. This is not to downgrade the city station, for through the years, it has done things which a commercial station could not do. For instance, it has been able to devote hours and hours to the continuous coverage of the United Nations sessions; a great service to a quite small percentage of the population, but, a very important one.

WQXR started in its first year to put on some of the best talent available. My diary for 1936 says that on November 15, we broadcast our first important "live" program. It was a recital by Roman Totenberg, a rising young violinist at that time who has since acquired a great and well-deserved reputation in this country and in Europe.

In that same period, we presented a series of programs called "Melody Through the Ages" which featured Roy Harris, probably the best known American composer of serious music of that time whose symphonic and other compositions have since become standard in orchestral repertoire. Roy Harris was looked upon as an *avant garde* composer in the thirties, and we staged a joint appearance of Harris and the late Olin Downes, music critic of the *New York Times*. The broadcast, planned as a discussion of the new idioms in music, became so heated that, at one time, it looked as if there might be physical violence, especially when Harris pointed his finger at the critic and said, "Mr. Downes, when did you ever attend a concert of contemporary music?". So-called "modern" music was broadcast so infrequently that even the knowledgeable music-lover was not accustomed to sounds which today are widely performed. WQXR's policy was to give people a chance to hear modern music but in moderation, for we knew that by giving it in small doses, the "patient" could more readily absorb it.

The first "live talent" sponsored series which we aired was a joint advertising campaign of Knabe pianos and Stromberg-Carlson radios. Each program presented a soloist which gave the listener a chance to hear a Knabe piano with optium high-fidelity transmission. This demonstrated Knabe quality and, similarly, Stromberg-Carlson could advertise its receivers as being capable of reproducing the sound of the piano as it should be heard.

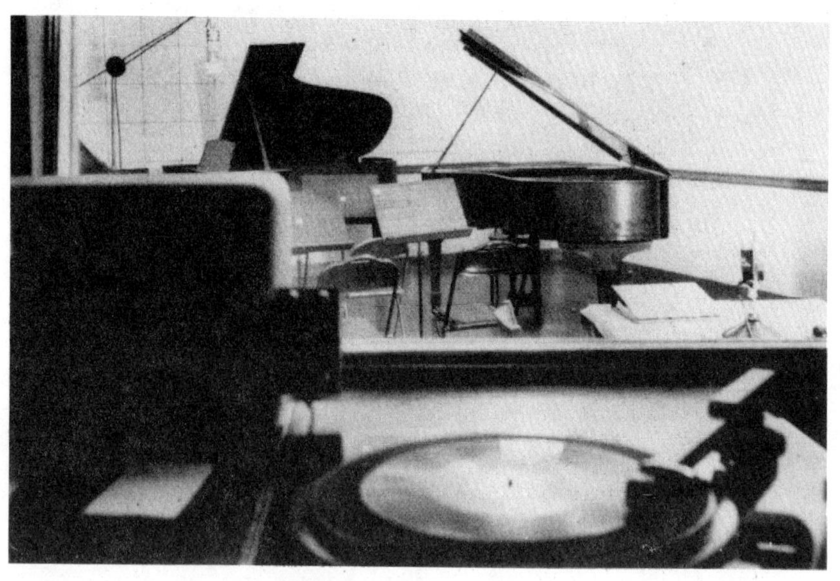

Studio set up for two-piano broadcast.

WQXR signed a contract with the Musicians Union early in our history, and the first man we hired was the pianist Jascha Zayde, a young and very talented performer. We immediately featured him in a half-hour program once a week called "The Development of Piano Music", one of our first efforts at painless musical education. Later, another pianist, Clifford Herzer, was added to the staff, and he and Zayde developed a two-piano team which became one of the drawing cards of our schedule. When Herzer went into the Navy, we were fortunate in putting together another team, Hambro and Zayde, who, for many years, occupied a feature spot on WQXR. It is gratifying to report that Jascha Zayde remained at WQXR for more than thirty years.

About this time, Eddy Brown, a famous violinist, came to WQXR as its music director. Through his wide experience as a concert artist, he was able to induce many excellent musicians to appear on the station. Most of them played without fees because of the publicity which an appearance on WQXR gave them, and the unions, fortunately, did not object to this at the time as long as we employed some union members on our staff. They also realized that the artist's reputation profited by a broadcast on WQXR.

Radio appearances such as these were possible only through the cooperation of the Musicians Union for the instrumentalists, and AFRA (now AFTRA) for the singers. Unfortunately for all concerned, the unions, after a while, prevented us from using these artists without fees which were entirely beyond our resources at a time when we were struggling to stay in business. The various unions soon organized all our staff, including announcers, engineers, musicians, writers and clerical workers. Much of the time and energies of the management were spent on union negotiations which, throughout the years, have always been a burdensome chore. There was no question of the staff's right to join unions. But the attitude of the unions, even when we were struggling to keep our heads above water, was curiously antagonistic to an enterprise which, if it succeeded, meant more employment for its members. I remember that in negotiating with one union, we pointed out that the conditions they were making might put us out of business. Their leaders replied that it was not their concern, and if we could not meet their demands, we should close up shop. Only the willingness of Jack Hogan, our friends and me to gamble on the future by putting up the money to meet their demands, kept us going until we got enough business to balance our budget—and that took a number of years.

During the period when the unions were more liberal about letting us broadcast live performances, we presented some interesting musical events. Among these were an opera, "The Frantic Physician," presented at the Juilliard School of Music by the graduate students and a concert from Carnegie Hall by the Women's Symphony Orchestra led by Antonia Brico, one of the few women orchestra conductors. These broadcasts were of more interest as novelties than as great music. Of much more significance was WQXR's exclusive broadcast of a preview of the New York World's Fair, May 1, 1938, a year before its official opening. The feature of the concert was the Beethoven Ninth Symphony performed by the New York Philharmonic, the Schola Cantorum, the Oratorio Society and soloists. This was not only an artistic event, but also historic in the realm of sound because it was probably the first high-fidelity transmission of a large group of instrumentalists and vocalists.

There were other events of this kind, but we knew we could not build a reputation for the best in music dependent upon the forbearance of the unions. By the early 1940's, because of new investors who had come into the company, we had more capital to work with, even though we were operating with losses in some months and profits in others.

By then we had a house group of ten instrumentalists under the direction of Eddy Brown. It performed small orchestral works, chamber music, and other music appropriate for an ensemble of that size. In addition, we engaged the Perole String Quartet, which had a fine reputation, to do a series for broadcast from our studios. Over the years we had constructed very adequate studio facilities at 730 Fifth Avenue.

In the summer, the great music event in New York was the nightly concerts by the New York Philharmonic attended by thousands in the big Lewisohn Stadium at the College of the City of New York on Washington Heights. Some of the concerts had been broadcast for a long time by WNYC, but finally we were able to arrange to have WQXR carry one concert each week throughout the Stadium season. These were distinguished concerts, and many great soloists appeared with the Philharmonic. The audience at home enjoyed the high-fidelity transmission of the music.

One of the high points of each season was the appearance of Mischa Elman as violin soloist. There was no artist who was more popular with New York music-lovers than this great virtuoso. One year Elman's manager informed us that we could not broadcast his performance unless we paid his regular fee—much larger than we could afford, so we regretfully told him that we would have to omit his section of the concert. If I remember correctly, he was to play the Mendelsohn Violin Concerto, and we decided that when it was Elman's time to play, we would inform our listeners that we had not been able to arrange to broadcast his performance and that we would play the same concerto on a record by Heifetz. When the weekly *Variety* heard what WQXR had done, it ran a story with a big headline reading, "Heifetz Fiddles While Elman Burns".

The best concert series of chamber music in New York at that time was the Sunday afternoon recitals in Town Hall of the New Friends of Music which had been organized by an active music-lover, Ira Hirschman. We contracted with him for the rights to broadcast these concerts and, for several years, they were enthusiastically received by chamber music lovers. This was another instance when we realized that to be entitled to be "America's No. 1 Fine Music Station", we had to satisfy minority tastes as well. These concerts brought to the microphones some of the greatest names in chamber music. For example, in the month of November 1942, the programs included the Coolidge Quartet, the famous two-piano team of Luboschutz and Nemenoff, Sacha Jacobson, the violinist, Ralph Kirkpatrick, the harpsichordist, the Musical Art Quartet, the world-famous Budapest Quartet and violinist virtuoso, Joseph Szigeti.

In this war period, we also put on special live concerts for the Treasury Department to encourage the sale of War Savings Bonds. Many famous artists volunteered their services for the series which was broadcast Sunday afternoons from our studios, with the composer, Deems Taylor, as commentator. In this same month of November, the Perole String Quartet was featured with additional guest artists, including George Szell, William Primrose, Benny Goodman, Adolf Busch and Rudolf Serkin.

The events I have described were some of the special gems of our schedule, and throughout our history, we were always able to bring unusual musical events to our listeners, including the first radio performance ever given by Vladimir Horowitz as well as the sensational appearance in 1958 of Van Cliburn for his first concert immediately upon his return from winning the Soviet Tchaikovsky Award in Moscow. More about these broadcasts later.

On the national music stage, the two prestige radio programs were the Saturday afternoon performances of the Metropolitan Opera and the Sunday afternoon concerts of the New York Philharmonic, both of which naturally drew

A World War II WQXR concert at Hunter College, New York. Tickets were given to those who bought Victory Bonds.

many of our regular listeners away from WQXR. We met this competition as best we could by programming operas opposite the Philharmonic and important symphonic recorded concerts against the Met. In this way, we attracted those who preferred opera to symphonic music and vice versa. These two series were, undoubtedly, the strongest music competition which WQXR had but, fortunately, it was for only a few hours a week during the opera and concert season.

We realized that we needed some important prestige orchestra to enhance our position in the music world. We tried to obtain the rights to the Opera and the Philharmonic, but that was impossible for many years, although we finally were able to get both. The Columbia Broadcasting System dropped the Philharmonic many years ago, and there was no broadcasting of that orchestra until the season of 1966-67 when we obtained the rights. People hailed this feature in our schedule, particularly because it was available to them on both AM and FM direct from Philharmonic Hall in Lincoln Center. Unfortunately, the station has been unable to carry these concerts since the 1966-67 season because the members of the orchestra demanded such an exorbitant increase in fees that we could not afford it nor could any other station. Thus the public was deprived of this great orchestra on the air.

In 1966, station WOR in New York decided that the Metropolitan Opera did not fit into its programming on the FM side and offered the FM broadcasts to us, which we were glad to have. So, now in the New York area, the listener in his car may hear the Opera on WOR-AM and at home over WQXR-FM.

Before we were able to capture these two prizes, WQXR had succeeded in broadcasting regularly the concerts of some of the nation's best orchestras. The first of these was the Boston Symphony Orchestra. In 1957, we made an agreement for a weekly broadcast of this famous organization direct from its Symphony Hall in Boston and its summer series from the Berkshire Festival at Tanglewood near Lenox, Massachusetts. Under Charles Munch and later under the baton of Erich Leinsdorf, WQXR has offered some of the world's finest music performed by the orchestra and outstanding soloists of the concert world. It was very expensive to send an engineering crew and announcers to Boston and Tanglewood every week, plus the cost of high-fidelity telephone lines from those places to our studios but we did so for several years. Fortunately, by this time the art and science of recording stereophonically on magnetic tape had reached such perfection that it was all but impossible to distinguish performances on tape from the original. The tapes made in Symphony Hall and at Tanglewood reproduced the sounds perfectly, (unfortunately even the coughs in the audience) so that one could not tell the difference although we always made it clear that the performance was on tape.

The success of the Boston series, particularly after we had proved the reliability and reproduction qualities of tape transmission, led us to add other important orchestras including the Philadelphia under Eugene Ormandy, the Chicago conducted by Jean Martinon, the Boston Pops with Arthur Fiedler, the Cleveland Orchestra under the baton og George Szell, and the Pittsburgh Symphony led by William Steinberg. Some seasons we had as many as three or four of these great American orchestras on our schedule every week, including the New York Philharmonic led by Leonard Bernstein. These, plus the Metropolitan Opera, left little question as to WQXR's position as the most important source of great music on the air. In addition, in the 1967-68 season, WQXR added two "live" orchestral series from Carnegie Hall and Philharmonic Hall. The American Symphony Orchestra led by the dean of American conductors, Leopold Stokowski and the National Symphony Orchestra of Washington, D.C. led by Howard Mitchell.

Important as these major musical acquisitions were to the standing and prestige of the station, the basic material which built our audience was the day-to-day broadcasts of carefully planned recorded programs. They were the bread and butter which nourished the musical appetites of our audience. That is what they wanted for musical sustenance.

Although this music was on records or tapes, our programmers were trained to regard the content of each disc as seriously as if the music were coming direct from a concert hall. They were told that each program had to be planned with the same care that a conductor gives to the design and balance of his public appearances. It was not simply a question of going into the record library of the station and picking records at random. Every recorded program had to have an artistic, historic or topical reason for the selections which were played. The large and well-trained staff which we had for this purpose, created a special image of WQXR which was difficult to imitate.

The ancestor of the important symphonic hours was the Wanamaker program in our first year. Later this developed into what has become one of the trademarks of WQXR—Symphony Hall, which has been heard every weekday evening from eight to nine o'clock for many, many years. It has become so well-known that the title has been copied by dozens of stations throughout the United States for their principal program of good music. Its signature—the chimes of London's Big Ben—has marked the hour of 8 P.M. for hundreds of thousands of people. In fact, some years ago the record of the chimes was broken just before it was to be broadcast, and there was no time to get another copy from our files so Symphony Hall went on the air minus its signature. The next day we received a telephone call from a harassed mother who said that the chimes had been the signal for her young son to go to bed, but when he did not hear their familiar sound, there was a family crisis getting him to go to sleep.

Back of our programs was WQXR's great record library which has become probably the most complete collection of records of good music in the country, with the possible exception of the Library of Congress. At first, the collection consisted of 78s, but these were later replaced by LPs and then by stereo recordings. Part of the collection which runs to about 75,000 discs is of historic musical importance, and such recordings are retained regardless of their age if they are usable for historical broadcasts. In addition, in recent years, the library is adding to its collection of tapes for, in most cases, they are superior to anything obtainable on discs.

"Symphony Hall" is the only program mentioned at this juncture because of its reputation, but another chapter will cover some of the other recorded programs which gave distinction to our schedules.

13

Basic Programming

WQXR goes on the air at 6 A.M. (7 A.M. on Sunday) and signs off at 1 A.M. This amounts to about 7,000 hours a year, and when you consider that WQXR has special music standards to maintain, it becomes a formidable problem.

How do you program so many hours a week? Aside from news and commentary and the feature musical programs I have already mentioned, the basic product is the result of the proper use of records.

Someone once said to me that the output of WQXR was excellent turkey—sometimes roast turkey, sometimes creamed turkey, sometimes just the drumsticks, sometimes turkey hash and finally good turkey soup. In a sense, that is all true, but whenever the turkey comes into your home, it has to be tasteful, well-seasoned and just the right dish for the time of day you are ready to enjoy it. It must never be what Broadway would call a "turkey."

The world of records is so vast that there is an almost unlimited supply, and by keeping up with everything that is released in this country and abroad, WQXR has been able to use this great supply of music to please almost every taste among lovers of good music.

Certain periods are devoted to symphonic works of an hour or more in length. Sometimes these programs have famous soloists playing a concerto, sometimes they include a large choral group just as such concerts might in Carnegie Hall or Lincoln Center. For opera lovers, there must be periods devoted to complete operas. For some programs, WQXR concentrates on chamber music with famous quartet ensembles, great soloists performing sonatas, song cycles and the like. On the lighter side, full renditions of Broadway shows and European operettas are always popular.

This variety in programming, properly balanced, has been the main magnet to attract the WQXR audience. But we do more than just play the music. We do not rely upon the announcer to read something from the jacket of the record. For many programs the announcer reads program notes about the composer or composition or both before presenting the music, and this writing is done by competent people who do not content themselves with the usual clichés about the number of children Johann Sebastian Bach had fathered and so on.

So that the scripts will add to rather than diminish the enjoyment of the concert, brevity is the rule. Some listeners want to know as much as possible about the music while others are impatient or more knowledgeable, so it is essential to satisfy these two tastes by just the right amount of talk. We cannot please them always, but we try our best.

Another essential in these programs is to have announcers who read with understanding. WQXR has always been proud of its announcing staff, and the audience, I think, has felt the same way about it. To pick announcers who can do all the things we require is not easy. An announcer for WQXR must, first of all, have a good voice and clear diction without affectation or speech mannerisms. He must sound cultured, but with no accent which is unfamiliar to standard American speech. In addition, he must be able to pronounce foreign names and foreign words with accuracy and authority. Because of the nature of our broadcasts, in addition to English he must be ready with acceptable pronunciations of French, German, Portuguese, Italian, Spanish, Russian as well as less frequently heard Greek and Hebrew. Our announcers do not have command of all these tongues, but they usually are fluent in one or two, and they have a special linguisitic talent for making them all ring true. The same high standard is applied to their broadcasting news where many unusual names and foreign words appear so frequently. However, there are certain exceptions which must be remembered: we do not want foreign pronunciation of names of familiar places. For example, Paris is not "Paree" nor is Berlin "Berleen."

To top off all these requirements for the job, the announcer must also be able to deliver commercial messages with conviction and "sell," yet he must not shout like a barker nor coo like a dove to make the appeal effective. It is not easy to find such qualifications and we have been known to audition as many as 200 candidates to fill a vacancy. Applicants who would be welcome at other stations do not necessarily fit our needs. On the other hand, some of the WQXR announcers would not be able to fill jobs at other stations.

As an example of what a candidate for announcer goes through as part of an audition, the following soon demonstrates whether he can handle WQXR programs:

ANNOUNCERS' AUDITION

Good evening, ladies and gentlemen. Symphony Hall tonight presents the incomparable Boston Symphony Orchestra, conducted by Dr. Serge Koussevitzky.

Before we announce this evening's program, we would like to take a few moments of your time to tell you of some of the composers and compositions which will be featured during the coming season. The Russian School of music will be well represented by the works of Dimitri Shostakovich, Igor Stravinsky, Alexander Borodin, Modeste Moussorgsky and Mili Balakireff. German music will be represented by the works of Johann Sebastian Bach, Dietrich Buxtehude, Ludwig van Beethoven, Arnold Schoenberg, Richard Wagner and Engelbert Humperdinck.

The following French masters will be heard: Maurice Ravel, Charles Marie Widor, Gabriel Pierné, Camille Saint-Saëns, Henri Vieuxtemps, and Jules Massenet. From the prolific Italian library of musical masterpieces we will hear works by Ermanno Wolf-Ferrari, Ottorino Respighi, Gioacchino Rossini, Ruggiero Leoncavallo and Umberto Giordano.

Dr. Koussevitzky will also conduct the orchestra in compositions by the following Spanish masters: Isaac Albeniz, Joaquin Turina, Manuel de Falla and Enrique Feranandes-Arbos.

This evening's program will commence with Claude Debussy's "L'Apres-midi d'Un Faune"; followed by the Mozart Serenade; "Eine Kleine Nachtmusik"; continuing with the symphonic suite of Rimsky-Korsakoff, "Scheherazade." The program will conclude with three Wagnerian excerpts: 1. Prelude from "Lohengrin"; 2. Daybreak from "Götterdämmerung," and 3. The "Siegfried Idyll."

In addition to our staff announcers, it was always part of our planning to have specialists in various fields of music conduct their own programs, write their own scripts, subject to the station's supervision, thus giving our listeners the benefit of experts' points of view and special knowledge.

One of the earliest of these ventures was in music written for the ballet and ballet choreographed to fit established music. To conduct this series, which ran for several years, we were fortunate to have the late Irving Deakin who, though not a dancer himself, was associated with ballet companies for many years and was the author of several authoritative books on the ballet. In the late 1930's and early 1940's, ballet was not as popular nor as generally appreciated as it is these days. Deakin's program undoubtedly brought to many people a better understanding of this art form because his word pictures of the beauty of ballet were almost as vivid as present-day color television.

Old records of by-gone artists have an attraction for music lovers. The reputation of singers and instrumentalists of the past create a desire to hear how they actually sounded, and there have always been collectors of these recorded rarities. Even though the accoustic quality and sound fidelity of these old discs leaves much to be desired, people want to hear what Sembrich or Schumann-Heinck or Ysaye or Paderewski, Caruso and many others sounded like, and most collectors are glad to broadcast their treasures to the delight of many in the audience. Various series of these historic programs have been presented on WQXR over the years. Much of this historic material is from WQXR's own collection.

Although American folk music is all the rage today, it was unfamiliar in 1938 when we had programs of records and live artists featuring John and Lucy Allison who were among the few at that time who were familiar with the folk aspect of American music.

One of the people who did much to popularize good music was Sigmund Spaeth. Those of you who are old enough will remember his famous "Tune Detective" demonstrations which showed how a contemporary hit "Yes, We Have No Bananas" was a direct descendant of Bach. He could do the same analysis with almost any tune you mentioned, and would trace its historical ancestry on the piano with appropriate humming. He did several series for WQXR, one of which was called "Dr. Sigmund Spaeth and His Record Library". He was always a performer who drew large audiences. In addition to his vast knowledge of music, his dynamic personality highlighted his broadcasts. Through his programs, he inspired many people with a love of the best in music because he took music off its pedestal and humanized it for those who had regarded it with awe.

When Town Hall in New York was a center of cultural activities, one of its most popular lecture series was the Layman's Music Course given by Olga Samaroff-Stokowski. An experienced concert pianist, she also had wide knowledge of music plus a charming personality. We engaged her to give a similar course over WQXR using records to illustrate her talks. This did a great deal to make our listeners more appreciative and knowledgeable about the masterpieces they were hearing every day over the station.

In complete contrast to those programs which involved a personality with recordings, we developed a daily broadcast which played music without the human voice being heard. Nothing was said—not even titles or composers. At this time, Mrs. Eleanor Roosevelt had written a newspaper column called "Just Babies." That gave us the idea to christen ours "Just Music", and it clicked right away and was heard from midnight to 1 A.M. as a soothing finale to our broadcast day. To the curious, we issued a list of the titles for a whole month in advance and mailed it to those who requested it.

The longest run of any of the specialized musical shows using expert personalities was "Nights in Latin America". In 1943, Evans Clark who was then head of the Twentieth Century Fund, had a hobby of collecting records of Latin American music. Although he was not a professional broadcaster, he offered to do a weekly program of these records which he had collected all over Central and South America. He had a fund of information about the lives and customs of our neighbors to the south, and he talked about them and played the native music. This was a program much different from the Latin rhythms used in popular dance music, and he opened a whole new world to WQXR listeners. Evans Clark appeared on the program for a while and then asked us to have it taken over by his sister-in-law, Pru Devon, who continued the series with skill and charm from then right to recent years. She made many trips to Central and South America and added rarities to the collection and gathered more material about the people of those countries. Although this program was a far cry from the classics and the Three B's, it always had a large following, which illustrates that most real music lovers like all kinds of music if it is the best in its class.

The success of this departure from the classics led us in after years to include a weekly broadcast of jazz. We felt that although jazz had been vulgarized by some, it was recognized all over the world as America's genuine contribution to music. We put this program under the guidance of John Wilson, an authority on the subject and the jazz music critic of the *New York Times*. Some of our audience appeared outraged at the sound of jazz in the sacred precincts of Mozart and Beethoven, but we felt (though some of us agreed with the objectors) that we could not ignore the artistic accomplishments of these contemporary American composers and virtuoso performers. On this same subject, long before we had a jazz program, we played the music of George Gershwin. This too caused some of our audience to criticize us in the 1930's, when his "Rhapsody in Blue" and his Piano Concerto were first heard on WQXR. We were sure that they were classics in a new idiom, and time has proved that we were right.

All of a sudden, the fad for "disc jockeys" hit the pop music stations. Each of them had his own fans, and the rivalry between stations and personality vs. personality became more violent every week. As their popularity grew, their salaries hit astronomical heights for those days. All this seemed far removed from WQXR yet we saw that we were in need of more personal rapport between our announcers and our listeners.

One day a youngish man came in to see my wife to find out if there was a job for him. His name was Jacques Fray,

Jacques Fray—the "Disque Jacquey"

and although Eleanor had never met him, she knew of him as one half of the fine two-piano team of Fray and Bragiotti. That team, in the earlier days of broadcasting, was famous for its skillful performances of good popular music. Both men had been friends of George Gershwin who brought them to this country from France, and their programs had done much to popularize his music.

Jacques had a charming French accent to his very good command of English and, in addition, his voice had sex appeal which Eleanor evidently spotted. She asked me to come to her office, and there I met him for the first time. She said that she thought he might make a good classical disc jockety, and we decided then and there to audition him.

The audition impressed all of us, and we agreed to put him on in the afternoons and publicize him as the first "classical" disc jockey. The publicity people could not resist referring to him as "Disque Jacquey." He was no learned authority on music although he knew a lot about it, but he had a style of presenting good music which was most pleasant and which endeared him to our audience for many years.

His French accent appealed to many, although it annoyed a few who referred to him as "that phony Frenchman." But Jacques was no phony. He was a native Parisian of an old French family, and I myself have seen a portrait of Jacques at about the age of three, painted by Renoir. This picture is in the famous Renoir collection at the Clark Art Institute in Williamstown, Massachusetts.

For more than 15 years, his program, "Listening with Jacques Fray" was daily routine for thousands of people who welcomed him into their homes or cars every weekday afternoon. It was therefore a great shock to all his associates and friends at WQXR, as well as to the audience, when Jacques suddenly died in January of 1963. Telephone calls, letters and telegrams expressed to us the sympathy of his listeners. People from all over, including Governor Rockefeller, wanted us to know how much he would be missed. We broadcast a memorial program and his former partner, Mario Bragiotti, talked about him and I said a few words in a last tribute to one of the personalities who had done so much for the station.

After a few weeks, we were fortunate enough to get Bragiotti to take his spot on our schedule, and he continued with the program for several years.

The classical disc jockey formula was proving popular with the audience so we explored the field. On a trip to London, I met some people who were signing up talent for radio appearances outside Great Britain where, of course, there was no commercial radio. I said that WQXR would

A recording session with Sir Thomas Beecham.

be interested in a series by the distinguished British conductor and raconteur, Sir Thomas Beecham. Sir Thomas, in addition to his great knowledge of music and conducting, had a reputation for biting repartee and a quick sense of humor. In 1949, we arranged with his London agent to record some sample programs and send them over for our approval. Sir Thomas was to select music recorded with orchestras he had conducted, was to say anything he wanted about the music and the people involved (within the bounds of propriety and with due consideration of the laws of libel).

The records arrived, and we listened to them expectantly. The program committee unanimously agreed that they were terribly dull and that the maestro needed some radio direction. When I wrote this to our friends in England, they replied that I knew how difficult Sir Thomas could be and that any attempt to tell him what to do would raise a storm of temper. However, they wrote that Sir Thomas was coming to the United States in about a month, and we could talk it over with him. The great man arrived, and we made an appointment with him to spend a full day recording and see what we could turn out. I decided that I would beard him in his den so I undertook the risk of being eaten alive by the British lion. We met, we talked things over, I gave him an idea of what the American public in general and WQXR in particular was looking for. We went over scripts together, and he was most affable and receptive to any suggestions I made. We then cut some records, and he agreed that they were much more entertaining and more reflective of his personality. He could not have been more friendly nor more willing to take suggestions. My conclusion was that the producer in England, knowing Sir Thomas' reputation as a curmudgeon, was frightened. The series was completed by more recording sessions in New York and London and went on the air at WQXR under the sponsorship of Hovis Bread, the English baker whose signs "Hovis and Tea" greet you all over the British Isles.

One series led to another, and we found more musical people who were good on the air. Deems Taylor was one of these. He was a leading American composer of opera and concert music at the time, and his experience as a music critic and commentator for network broadcasts made him particularly suitable for doing programs for us. His knowledge and his subtle humor resulted in a first-class series.

We also discovered a fine commentator right in our own music department. He was a young fellow named Fred Grunfeld who had a rare combination of talents. He had an extensive knowledge of the musical world, past and present, and had great ability as a script writer and an intimate approach on the air. We gave him the assignment of writing what we called "Music Magazine" for an hour five times a week in top evening time. The idea behind "Music Magazine" was that each program should deal with some subject or personality, contemporary or historical, which would be illustrated by music and with commentary which would enlighten as well as entertain the listener. It was not to be technical but broad in its appeal, similar to a magazine article. Fred Grunfeld had a special gift for doing this program. Our main problem with him was to get his script finished on time. I can recall seeing him tear the last sheet out of the typewriter just in time to walk into the studio and go on the air.

The bicentennial of the birth of Mozart occurred in 1956, and we decided that this was the time to do a series of special programs to illustrate the greatness and wide range of Mozart's genius. We chose an experienced musicologist, Herbert Weinstock, to write and present a weekly program for 26 weeks so that our audience would have a full panorama of Mozart's works. In addition to this more or less educational series of broadcasts we, of course, included in our regular schedule many Mozart symphonies, operas, chamber music and smaller works.

Arthur Sulzberger, publisher of the *Times* and my boss, always thought that we played too much Mozart, Haydn and others of that period whose music he did not enjoy. This Mozart bicentennial year on WQXR was a trial for him, and he frequently complained to me. When I asked him if we should discontinue it, he would say, "Oh no! You're running the station."

Sometimes these encounters took the form of a good-humored exchange of "poetry". One which I have preserved is this:

OWED TO A DISCONSOLATE LISTENER
(who pays the bills)

Musicians say that none promotes art
Quite like that fellow Wolfgang Mozart
But I don't like him much, nor Haydn
If asked, I'd have a job decidin'
Which one I'd rather quick turn off (sky)
To listen to my friend Tchaikowsky.
Give me more Strauss, Sibelius, Schumann
But please go lightly on some new man
Who seeks for discords, vainly striving
To get his list'ners by conniving
To make them shudder, head to sandal
I really think I'd prefer Handel.
Cut out Debussy, give me Gershwin
Among elite, I'll ne'er a purse win
But I like melody with music
(My taste must certainly make *you* sick).
Give me some Liszt, Puccini, Lehar
And add a stirring bit by Elgar.
Let's have some pomp and circumstances
And melody of old romances
Let's ban each harp or organ solo
And as for volume—when it's so low
That you can really never hear it
Then turn the knob if you are near it
You'll save all that electric power
Between the news that marks the hour.
A. H. S.

To which I replied:

WEEK-END THOUGHTS ON PAYING THE PIPER

You claim that you are no musician
But yet at verse you're some technician.
I cannot hope to take much pride in
These rhymes defending Papa Haydn.
As you may guess, it's quite a job
To please all listeners, smart or snob.
The things on which some vent their hate
Are viands rare on 'nother's plate.

Ravel, Debussy, Suk and d'Indy
Are thought by some as simply dandy
Another says it pains his belly
Whenever we play Locatelli.
"Why play Tchaikovsky in the morning?
It's Bach we want when day is dawning."
"Why Bach with sizzling breakfast bacon?
Tchaikowsky's needed when we waken."
"At noontime who wants Mister Mahler?
He's not for listening in the parlor."
"At supper I resent your Brahms
For me let Elgar weave his charms."
Say some: "At midnight make it dreamy"
Others: "I like it hot and steamy."
From this you see I cannot win.
What hope is there for the fix I'm in?
E. M. S.

When I retired in 1967, Arthur Sulzberger wrote me a warm and friendly letter, the last paragraph of which recalled our perennial "battle." It read: "Even though you won't be on Forty-third Street at your daily duties, do you think we could find some way to keep alive our Eternal Triangle—Sanger, Sulzberger and Mozart? I would miss that never-ending debate!"

But it did not end there. When Mr. Sulzberger died in December 1968, at a memorial service, James Reston, then Executive Editor of the *Times*, was the only speaker at the simple ceremony. Mr. Reston, talking about the kind of service Mr. Sulzberger wanted, said: "Five years ago, when he was already seriously ill, he wrote out the instructions for this occasion.

"There were to be none of the ghoulish trappings of death, he said. No fancy casket, no mountains of flowers smelling of the grave and on pain of eternal punishment—

for some reason not quite clear to me—no Mozart. He had a thing about Mozart and was forever chasing the poor man off the *Times* radio station!"

Arthur Sulzberger had the last word!

14

Inside WQXR

In the course of more than 30 years a growing organization attracts many kinds of personalities and temperaments. This was particularly true of WQXR because it was a pioneer and something new in "show business." The people who came and went, especially those who were with us for a long time, all contributed to the character and growth of the station in one way or another.

When Jack Hogan and I got together to establish WQXR, Jack already had a young man who chose the music for the programs, operated the turntables and control panel and simultaneously announced the program on the air, reading from continuity he had previously written. The name of this all 'round utility man was Douglass MacKinnon. He had a fine musical background as a result of many years of study. Above all, he had good taste in music and understood our policy. This does not mean that we always agreed with his choice of music, but he knew much more about music, in a technical sense, than we did. For instance, I remember that we frequently complained to him that he was playing too much Delius. In 1936, the music of Delius was not familiar to most of our audience who thought it was "way out." The growth in respect and admiration for that British composer in the intervening years shows that Doug MacKinnon was right.

Because MacKinnon was a sort of one-man broadcasting station, he was most valuable to us at the start. He was an individualist, and as the station grew in those first years and more people were needed in our operations, he could not accustom himself to working in an organization. He wanted to work at times convenient to him without regard to the fact that other people needed his cooperation on a regular schedule to get our programs on the air. We urged him to fit into the organization, but he could or would not do it and did not understand that as we grew we were becoming a business and had to operate in a systematic way. Finally, after a year or two, we had to part company despite his very real abilities.

By that time, we had hired Eddy Brown as music director. Brown had been an infant prodigy violinist, and he had given recitals in many parts of the world. As a pupil of the famous Leopold Auer, he was rated one of the fine violinists of the day. He also was well known to radio listeners because he had broadcast many concerts on WOR, New York. As a young boy, Eddy Brown had played before King Edward VII of England who gave the youngster a gold watch in tribute to his performance.

Brown's chief contribution to the WQXR of this period was the development of a staff string orchestra. This group of ten players broadcast several programs a week with Eddy Brown as conductor and sometimes soloist. The men he had selected were experienced musicians and the quality of programs produced by them was commendable. Brown also played solo programs, and he frequently did some of the great works of the sonata repertoire with well known pianists. Sometimes the group was divided into quartets and quintets to play chamber music. It was a novel experience for the New York music-lover to hear so much fine live music on the radio.

As Eddy Brown became more involved in administrative duties, we found it necessary to find someone with wide conducting experience to develop the group further for, after all, Eddy Brown was a violinist rather than a conductor. We brought in Leon Barzin, an experienced orchestral player and, for many years, the conductor of the National Orchestral Association Orchestra. This was a symphony orchestra designed to give orchestral training and experience to young men and women. Over the years, it had developed some fine first-desk players for symphony orchestras. Barzin was able to continue this work at the same time that he took over the musical direction of WQXR. As Brown became less active and later left us to resume teaching, Barzin developed live groups for our broadcasts. Out of this unit of ten players, Barzin created a chamber music group, a two-piano team, soloists, salon music and had special arrangements made to perform classics of the string orchestra literature.

Leon Barzin was with WQXR until the time came when the cost of maintaining the full group was so great that we could no longer afford it, and we were forced to limit our live music to quartets and quintets. Barzin moved to Paris where he now lives part of the year. He conducts in Paris and is also a guest conductor in other European cities. He still is music director of the National Orchestral Association and returns to New York each season.

As we were forced to reduce the number of live performances, we began to concentrate on the creation of a wider and more original use of records. Increasing hours in our daily schedule meant the planning and preparation of more hours of fine music. This required a staff of eight or ten musically knowledgeable people. My wife Eleanor had general supervision of this personnel, and under her direction, the broad programming plans were decided upon. Working out the details and writing the continuity were the duties of the staff under her.

In 1943 we added to our staff a "music adviser" in the person of Abram Chasins. He, too, had a distinguished virtuoso background as a concert pianist and was also a composer, many of his works being already in the piano repertoire. From early childhood, he had shown great talent as a pianist and had attended the Curtis Institute in Philadelphia where he was a pupil of Josef Hoffman, probably the greatest pianist of that era and the director of the Curtis school. Chasins became a member of its teaching staff.

Abram Chasins did a great deal for WQXR over the period of more than 20 years that he was with us. First as a con-

Yehudi Menuhin, the violin virtuoso, about to be interviewed.

Jascha Heifetz and Abram Chasins talking to a young contestant in WQXR's series "Musical Talent in Our Schools."

sultant and later as music director, many of his ideas added to the attractions of the station. He knew so many people in the world of music that his contacts brought many performers to our microphones. In addition, great artists such as Heifetz, Horowitz, Rubinstein, Kreisler, Serkin, Menuhin, Leonard Rose, Francescatti, Mischa Elman, Van Cliburn and many others appeared for interviews on the air with Chasins. Some served as judges to select the best talent among the young people who competed in our project, "Musical Talent In Our Schools," which Chasins supervised and which was an educational project sponsored by the *New York Times.* These musical personalities were willing to help because the contest promised to discover new talents (which it actually did).

For many years, Chasins was responsible for a series of solo recitals WQXR broadcast every week which presented some of the younger talent. Many of the great names of the concert stage today played on these broadcasts before these performers were as well-known as they are now.

One of the exciting events to which Chasins contributed much was the first radio broadcast ever made by Vladimir Horowitz which WQXR broadcast exclusively. The concert was broadcast from Carnegie Hall on April 23, 1951. The feature of the recital was Horowitz's own piano version of Moussorgsky's "Pictures at an Exhibition." It was a magnificent performance, and he followed it with several encores, winding up with his special arrangement of "The Stars and Stripes Forever" by Sousa which many critics agree could not be played so brilliantly by any other pianist. The response to the broadcast was the greatest we had ever experienced, and it appeared that everybody who loved music had listened. The excitement over this concert and broadcast was enhanced by the fact that this was Horowitz's last appearance before a sabbatical which was to last until 1953.

We had made a recording of the broadcast and a week later, we had a small luncheon for Horowitz after which we played the record. When he heard it, he was very pleased and acted like a happy child. Before we played it, I told him we had made a record, and he modestly asked, "How was it? Did I play well?"

Another great musical "beat" for WQXR took place seven years later. On April 14, 1958, the news from Moscow was that the winner of the famous Tchaikovsky Competition was an American—a young pianist from Kilgore, Texas named Van Cliburn. The fact that the Russians had given the first prize to an American was not only unexpected but exciting. The general public, even those who had no interest in music, became as enthusiastic as if an American had defeated a Russian for the world's heavy-weight boxing championship.

Van Cliburn was all over the newspapers and a current topic of conversation.

Cliburn was certainly not well-known to the musical public and, to some extent, not even among professional musicians. He had been taught as a child by his mother who was a piano teacher. He had entered the Leventritt Award contest in 1954. That competition had great prestige in the music world of the United States because its board of judges comprised some of the most famous musicians and performers of the day. In 1954, the tall young blond from Texas had competed and had deeply impressed the judges (one of whom was Chasins), and had won the prize. Cliburn had appeared as soloist with some of the important orchestras throughout the country, for that was the prize for winning the competition. Despite this, when the cables arrived saying that he had won the Tchaikovsky prize in Moscow, the name Cliburn did not ring a bell with many concertgoers.

Chasins had mentioned to me that he knew Cliburn from the Levintritt contest, and recalling the triumph of our Horowitz broadcast, I said, "Abram, let's get Cliburn's first concert when he returns to New York." He agreed that it was a fine idea and said he would try. We sent a cable to the New York *Times* bureau in Moscow asking them to get in touch with Cliburn and tell him what WQXR wanted. After several exchanges of cables between New York and Moscow, the arrangements were made to broadcast the triumphal concert from Carnegie Hall on May 26.

This was probably the largest expenditure ever made by the station for a single broadcast. In addition to the fees for the orchestra, we decided to promote the concert by advertisements in the newspapers to get a big audience. When Cliburn arrived in New York, there was a ticker tape parade up Broadway to City Hall. In addition to the City Hall welcome, there was a luncheon for him given by Mayor

Wagner and all the other honors generally associated with the return of a conquering hero, which indeed he was.

The weeks before and after the first concert were filled with all sorts of receptions and honors for Cliburn. Sharing the celebrations was Kiril Kondrashin who had conducted the Moscow State Symphony for Cliburn's final prize concert in Russia. He was in New York to conduct the orchestra for the big Carnegie Hall event which WQXR was to broadcast. The two of them, appearing together so many times, created the first sign in many years of an "entente cordiale" between the United States and the Soviet Union. A great friendship had developed between these two musicians which flourished despite the fact that Cliburn did not understand Russian, and Kondrashin could not speak English. While all the excitement was on, one day at lunch Arthur Sulzberger asked me how they communicate. I said, "They don't communicate; they just embrace!"

The concert was, of course, one of the highlights of the season. Carnegie Hall was filled to the top gallery with those who had rushed in time to buy tickets. In addition, the broadcast had probably the largest audience WQXR ever reached, for it brought Cliburn's sensational artistry to hundreds of thousands of people, many of whom, I venture to say, had never heard nor appreciated serious music before. The next day, in fact that very night, telegrams and telephone messages of thanks poured into the station to be followed by hundreds of letters in the next few days. It was one of the memorable events in the history of WQXR.

As the years went by, we realized that it was vital to train a younger generation to take the lead in developing the station both musically and in general administration. One of the crucial spots was the program department. We looked around for special talent, especially people who knew a lot about records and could write about them and even go on the air and talk about music and musical personalities.

Fortunately, in the middle 1950's, we found just such a person in Martin Bookspan. He was a musicologist who had been on the staff of the Boston Symphony Orchestra for some years, not as a performer, but as a spokesman for the orchestra on radio interviews as well as its commentator on broadcasts of the orchestra. We put him in charge of planning all our recorded programs.

When we started to air the Boston concerts, it was logical for us to make Marty Bookspan the commentator. He knew the orchestra, man by man, and had been around Charles Munch who had been the musical director of the Boston for many years. This venture was so well received that Marty became the voice of many other programs on WQXR. Today, Marty, though no longer a member of the WQXR staff, is a free lance commentator for some broadcasts. Now in the post of program director is one who, though young in years, has had long experience at WQXR. He is Robert Sherman who came out of the army right to us as a junior in the program department.

Bob has become one of the station's most successful personalities on the air. He broadcasts two hours every morning, Monday through Friday, his "Listening Room" program. He talks about the music the listeners are going to hear and almost every day he has musical personalities on the air with him. He and his guests talk about music and their special expertise. Frequently the guests perform "live" from the WQXR concert auditorium. People seem to like this because it is in the best tradition of WQXR—it offers good music informally, interestingly and informatively.

A few years before Bookspan joined the staff, we were still looking for a young man who could come in as an assistant to my wife and to me, for we felt that after almost 20 years, it was time to think of a younger generation to train for future top jobs in the organization. Eleanor and I had been away for a vacation, and on our return, Norman Mc-

Gee told me he had interviewed a young fellow who might fill the bill, and he asked me to talk to him. My diary indicates that I interviewed Walter Neiman on August 18, 1953. It goes on to say, "Seems just the sort of a boy (27) worth training."

We had had unfortunate experiences in the past trying to find an assistant. They had difficulty in adjusting to WQXR's unique policies because they had come from larger organizations where the whole atmosphere of the operation was different. Frequently, they were not accustomed to keeping their noses to the grindstone of our daily chores. Some were not ambitious enough, and some were too ambitious.

In hiring Walter Neiman we were picking a younger man than we had ever tried before, and one who was accustomed to a more personalized operation. We had also made the mistake in the past of hiring a person and making him an assistant right from the start. Walter came in more or less as a trainee, and we told him we wanted him to work in every department and become familiar with what made the wheels go 'round. He was to learn and not to administer, and this was the routine followed for a long time until we saw that he fit into the organization and that he had that special feeling and affection for WQXR which would make him valuable to us.

In 1960, the *Times* asked me whether I would like to go to Paris for a short time to help organize the International Edition of the *Times* which was just getting started in the French capital. Being a newspaper man at heart and having always wanted to run a paper, I accepted with the understanding with Orvil Dryfoos that it was to be limited to a few months until I got things running smoothly and then I was to return to WQXR. So Eleanor and I packed up and moved to Paris, and the management of WQXR was placed in the hands of Norman McGee and Walter Neiman.

The author and Eleanor—on the day of his retirement.

Fortunately, I was able to help create a smooth-running organization in Paris in about four months, and although Orvil urged me to stay a few months longer, I was getting homesick for my family in New York, for New York itself (although Eleanor and I always think of Paris as our second home), and we also missed WQXR. So we returned home and when we did, Eleanor decided that Walter had done so well in her absence, that now was the time to turn over her job to him. So he became Program Director in the middle of 1961 and thus reached the top echelon of the station.

In 1965, I became Chairman of the Board and the job of Executive Vice President and Chief Operating Officer of the station went to Norman McGee. At the same time, Walter Neiman was elected Vice President in charge of operations, and Bob Krieger was made Vice President in charge of sales. When my successor, Norman McGee, retired in 1968, Walter Neiman became Vice President and General Manager.

Because this chapter is about "inside WQXR," the real insides of the station must not be overlooked, although the average listener thinks about a radio station in terms of what he hears and not what goes on "behind the scenes". The strong engineering foundation of WQXR was, as I have mentioned earlier, the great pioneering ability of Jack Hogan and the operating and construction skills of our first Chief Engineer, Russell Valentine.

It was Jack Hogan's inspiration which was the basic idea for the creation of WQXR. His reputation as a pioneer in radio and his accomplishments in that field, caused the Federal Communications Commission to grant him permission to experiment in both television and high-fidelity radio as far back as 1934, which led directly to the idea of quality broadcasting which he and I were later able to launch.

For many years he and I were partners and I had ample opportunity to know this man of extraordinary talents. He was an unusual combination of engineering "know-how" and artistic standards. Like many scientists he was a dreamer, and his dreams had results many years after they originated in his laboratory. They included his single dial control for radio receivers, which I have mentioned before; high-fidelity transmission of sound which came into practical application with the advent of WQXR; his faith in Frequency Modulation (FM) and his cooperation with Edwin H. Armstrong, the inventor of the system; and facsimile broadcasting which demonstrated the feasibility of transmitting pictures and printed matter over the airwaves or over wires, a thing which is now a common method of rapid communication. There are various systems of facsimile transmission, but some of the principles laid down by Jack Hogan were fundamental.

The *New York Times* experimented with the idea of using the Hogan facsimile system to transmit a small newspaper right into the home. Using the transmitter of WQXR-FM, in February 1948 it put out six editions daily of a four-page newspaper, 8½ by 11 inches in size. The transmission was picked up by experimental receivers, made by General Electric, which were located at 14 department stores in the New York area and at the Columbia University School of Journalism. The joint experiment of Hogan and the *Times* ran for about a month.

At the convention of the National Association of Broadcasters in Chicago the following year, the Hogan facsimile system had its first demonstration to show broadcasters how their FM stations, as an auxiliary service, could send pictures, diagrams or any printed material into peoples' homes. To show this, a photograph was taken with a Polaroid camera. In one minute the picture was placed on a facsimile transmitter and sent by wire to WMAQ-FM atop the Civic Opera building and then broadcast by its FM transmitter without interfering with the sound program simultaneously on the air. The facsimile signal was picked up back at the Convention and the whole page, with the photograph, was reproduced in three minutes and 18 seconds. The entire process, from snapping the camera to receiving the printed page, took less than four and a half minutes.

Facsimile broadcasting to home receivers might have become universal had not the idea been launched approximately at the same time as television began to take the country by storm. Yet today facsimile systems are used all over to transmit, usually by wire, printed material and personal communication services.

As a person, Jack was an attractive companion and one who inspired confidence. He was one of the most systematic men I have ever known. As he sat at his desk puffing away at a pipe until the bowl was empty and immediately re-

John Vincent Lawless Hogan.

placing it with a cigar, he was always calm and thoughtful. If his secretary put through a telephone call to him, he would pick up a pad and pencil and make notes to indicate that Joe Jones had spoken to him at 10:32 A.M. and the gist of the conversation. At the end of the day he would enter all these jottings into his permanent diary. Everything except routine matters went in the diary for future reference and I know that these records were most useful when he testified as an expert in patent cases, for which he was frequently retained. He was purely business during the day, but in the evening he liked to relax with a drink and perhaps play and sing some of his favorite tunes if a piano was handy.

In the early days, his confidence in the future of WQXR never lagged and he backed up his confidence with money when it was needed even though he did not always have much to spare. There were days when we could not meet our Friday payroll until Jack gave us a check.

When the *Times* bought the station he received a five-year contract and at the end of that period he was retained as a consultant on engineering matters for some years until it was necessary for him to devote all his time to the development of his facsimile systems. He was active in his laboratory almost to the time of his much-lamented death in December 1960.

Under the guidance of Jack Hogan, Russ Valentine had done much for the technical operation of WQXR and when Val died at a quite early age in 1951, WQXR lost a very valuable member of the staff, one whose important services dated back to the time when the station was hardly more than a gleam in the eyes of Jack and me.

He had trained an assistant, Louis Kleinklaus, who was familiar with Val's methods and shared his insistence on quality engineering to match quality programs. It was therefore easy for us to appoint a successor. Kleinklaus had come to WQXR many years before as an engineer and, by his competence, worked up until he became second in command. Under his supervision, all the technical developments from 1951 to 1970 took place. These include the several years of concentrated work to get our 50-kilowatt AM transmitter built and on the air; the installation of the FM trans-

The fourth 300-foot tower of the AM 50-kilowatt transmitter at Maspeth, Long Island, being erected.

mitter atop the Empire State building, and the pioneering which the station did in stereo transmission.

Aside from the WQXR people engaged in programming, engineering and announcing and the "air personalities", there were major contributions made by those in the more mundane field of selling advertising. I have already mentioned Norman McGee and Bob Kreiger, but there was one man who was chief of the Sales Department before them.

It was Hugh Kendall Boice, known throughout the broadcasting industry as "Ken." He was one of the pioneers in broadcast advertising and had been a vice-president of the Columbia Broadcasting System in its earlier days. WQXR needed someone who could give stature to the selling effort, someone with great experience in the field, and with contacts with advertisers and advertising agencies. In 1940, we were fortunate in finding just the right person in Ken Boice. Not only was he the experienced person we needed, but he had the personality that fit the job perfectly. He took on the position of Vice President of Sales realizing that our selling policies and our resources were far different from CBS, but he accepted the challenge.

He gave WQXR the status it needed on Madison Avenue and, while doing so, helped our whole sales organization and our sales results. Ken Boice was a very fine person, and it was a privilege to have worked with him. When he became ill in 1950 and had to retire shortly afterward, he was a loss to our organization. Fortunately, by that time Norman McGee was ready to succeed him.

To the listener, the character and particular personality of a radio station is largely determined by the voices of the people he hears. I have already written about the difficulty of finding announcers with the right qualifications for our staff. Once we hired them, it seems that they became permanent fixtures in our operation. The backbone of our announcing staff today is the same men who have been with us for many years, one dating back to the first year, 1936. Others have been on the staff for 20 or 25 years. Fifteen years of saying "This is WQXR, the Radio Station of the *New York Times*" seems to be about the average service record.

Our announcers are, I believe, a breed unto themselves. They are intelligent, cultured, assertive and sometimes very difficult to get along with. At heart, every announcer is an actor and perhaps feels that he could have been another

John Barrymore if he had not been diverted to the airwaves. So sometimes he takes out his thwarted histrionic ambitions by making life difficult for the management. Most of them are strong union men, members of that very dynamic union, AFTRA (American Federation of Television and Radio Artists). Announcers are more difficult in contract negotiations than any other group I have had experience with. There were frequent threats of strikes by the announcers, but WQXR never had one.

The names of the "Big Ten," I am sure, will be familiar to those within listening range of WQXR, and I list them here in alphabetical order: Peter Allen, an erudite, serious-minded, soft-voiced member of the staff for a long time; George Edwards, whose cheerful manner so well fits the "Bright and Early" segment which has helped so many people start the day on the right foot; Melvin Elliott, with that deep, soulful voice which listeners have enjoyed ever since the end of World War II; Bill Gordon, whose personality reflects his skill as a singer as well as an announcer; Albert Grobe, our chief announcer, who until recent years voiced most of the daytime news reports of the New York *Times* and who has now retired; Bob Lewis. the man with the colloquial approach to all sorts of continuity; Lloyd Moss who, had he gone on the stage, would have been a matinee idol and, in fact, is one on the air; Duncan Pirnie who, if punning were a fatal disease, would have been dead years ago and who is undoubtedly our announcer who can do so many different kinds of programs with equal skill and who our listeners either adore or hate; Chester Santon, one of our veterans whose voice is familiar to thousands; and William Strauss ("Bill"), whom I hired as a very young man the first year we were functioning and whose ability has not dimmed with the years.

For many years, we tried to interest our audience in interview programs done by attractive intelligent women. The best known of these is Alma Dettinger who, every afternoon for a long time, talked to well-known people about books, food, politics, fashions and whatever was current news at the time. She was a warm and friendly person who had a fine following. Unfortunately, she became ill and the strenuous job of five broadcasts a week was too much for her, and she retired, much to the dismay of her many friends in and out of the station. After her, we put on another program called "Observation Point" featuring Miss Duncan MacDonald—yes, "Miss" is correct although she seems to be the only woman I have ever heard of to bear the masculine first name of Duncan. Her personality on the air was far from masculine, and she created a program of wide interest. Now, WQXR does not have this kind of program. The reason for this decision is that research over the years showed us that because of the musical emphasis of the station, most of our audience had little patience with talk except news and news commentary, and we lost too much audience at the times these programs were on. Although both these women had a great many fans—enthusiastic ones at that—there were not enough of them, and we had to replace their programs with good music.

15

Special WQXR Policies

BECAUSE WQXR was a special kind of radio station, we had certain policy problems which were not typical of the industry in general. Our strict supervision of advertising, especially our ban on singing commercials, caused more problems than anything else. It was very hard to tell an experienced advertiser who had a singing jingle that was a great success all over, that WQXR would not accept it. Some agencies and sponsors resented this, others respected us for our stand but did not give us the order; others gave us a chance to submit our own ideas and often accepted our version.

One of the latter was the United Fruit Company whose "Chicquita Banana" spot was a big hit on other stations but which we would not accept. But when he accepted the special commercial which WQXR created, the advertising manager of the company sent me a framed cartoon which showed an irate tycoon pounding his desk. Underneath was the caption, "There isn't any reason for it. . . . It's just our policy!!"

The biggest noise we ever stirred up in the broadcast advertising world was not because of something we would not accept, but something we solicited which other stations would not take—liquor advertising.

It is the practice of most radio stations, except some in "dry" areas, to accept advertising for beer or wine or both, but whiskey and other "hard" drinks are beyond the pale. For many years, we had adopted the same restrictions on hard liquor, but from our earliest days we accepted advertising for wines and beer and cordials. Cordials were not accepted by many stations but we did on the theory that drinking sweet liquers would probably make you ill before you became drunk. We went so far as to advertise vermouth even though the copy dealt with how you used it to make cocktails with gin or whiskey.

The question might be asked, why, if we were so careful about the products we accepted and the character of the advertising, did we take hard liquor commercials which most other stations rejected?

The reply is simple: WQXR appeals to an audience in the metropolitan area of New York. It is a sophisticated audience. New Yorkers are constantly exposed to alcoholic beverage advertising in their magazines, newspapers, billboards and in the subways and buses. Also, we knew a great deal about the special WQXR "type" who was even more worldly than most New Yorkers. The proof of that was evident from the many years in which we had advertised liquors (not including whiskey and gin). It had been received without protest and had, in fact, proved productive for our sponsors.

WQXR is a member of the National Association of Broadcasters which has a code of advertising standards that members are asked to subscribe to. This code's provisions for the length and amount of advertising within specific time periods is far more liberal than our own restrictions, and in all respects, save one, the entire code is far too liberal measured by our practice. The one provision which we could not live with was the prohibition against advertising of cordials, cocktail mixes—things which WQXR deemed appropriate. Of course, it barred whiskey, gin, vodka and the like, but that did not concern us, for those too were on our banned list. But because of the restrictions on other beverages, we did not subscribe to the code.

We had considered among ourselves many times the feasibility of WQXR letting down the bars (no pun intended) and accepting hard liquor advertising with certain restrictions, but we had never reached the point of deciding to do so. As the years went by, we felt more and more that it was a hypocritical position which the National Association of Broadcasters was taking. When one looked at the dozens of barroom scenes in Western pictures on TV, one wondered whether the viewer was supposed to think that the cowboys were drinking milk. The TV screen could show them in saloons, but it could not advertise whiskey. Another foolish rule in our opinion was one which said a person could raise a glass of beer to his lips to advertise a certain brand, but the TV screen could not show anyone actually drinking the stuff. As to the general rule of restricting hard liquor advertising on TV, it made some sense because of the great variation in mores and in age in the national TV audience, but it did not make sense to us who were serving an adult urban audience.

Finally, in March 1964, we decided to take the plunge. We announced that we would accept advertising for whiskey, gin, vodka and other hard liquor to be broadcast after 10:30 P.M. to avoid advertising to children—although our previous surveys indicated that only about 3 per cent of our total audience was under the age of 18. We also made it a rule that in the time period we were opening up, we would not use spot announcements but only sponsorship of programs of a half-hour or more duration.

Then the excitement started. LeRoy Collins, former governor of Florida and at that time president of the National Association of Broadcasters sent me a telegram, and later telephoned to me and urged us not to take this step. He agreed that the situation of WQXR was different than the average station, but he was sure our action would lead other stations with different audiences than ours to follow suit and thereby encourage Congress to step in and regulate all advertising on the air. I replied that the decision we had made was only after long and careful consideration and that the safeguards we were taking were adequate. I emphasized that the policy we were making applied to WQXR and very few other outlets and that the "self-regulation" which he advocated left it up to each station to decide for itself whether its audience was mature enough. Furthermore, we certainly did not advocate other broadcasters following our lead.

Howard H. Bell, director of the code authority of the NAB, said to the press that he thought the *Times* station was instituting an "unwise policy" that could give more ammunition to the dry forces in Congress. Naturally, the broadcasting industry as a whole did not want that to happen because it might eliminate the advertising of beer and wines which were an important source of revenue for many stations. We therefore received little or no support from the broadcasters.

One of our strongest supporters was the largest circulation daily in New York, *The Daily News*, which ran this editorial:

HURRAY FOR STATION WQXR

> . . . and its decision to carry whiskey commercials after 10:30 P.M.; and nuts, say we, to LeRoy Collins, National Association of Broadcasters chief, for his sanctimonious plea to WQXR to change its mind. Liquor exists, it is lawful, a lot of people use it, and liquor ads aimed primarily at adults are legitimate. Better drown your sorrow, Collins, in a large economy size Tom Collins.

Though hardly written in the style of the *New York Times, The Daily News* certainly summarized the matter succinctly.

Another supporter of the stand taken by WQXR was Brooklyn's veteran representative in Congress, Emanuel Celler. He telephoned me to congratulate us on our fight against what he called "Bible Belt hypocrisy" and said that he saw no logic in keeping liquor advertising off the air if it can be placed in print media.

But Congressman Celler's stand was not typical of Congress. A few days after our announcement, two top-ranking members of the Senate Commerce Committee introduced a bill to make it unlawful for radio and television stations to advertise hard liquors. Various Senators and Representatives took the opportunity to air their views, most of them opposing us. Others tied in the liquor issue with cigarettes and asked why the stations were at that time full of such advertising after the disclosures of the health hazards of smoking, yet were hypocritical enough to profess to believe that liquor was a greater menace.

Sign opposite Radio City Music Hall, Avenue of the Americas, during the liquor advertising controversy.

Batch of Congressional Bills Banning Booze Blurbs Expected If WQXR Stance Gains Momentum

Headline in *Variety* at the time of WQXR's acceptance of hard liquor advertising.

The fuss stirred up at that time about cigarettes continued to increase and resulted finally in prohibiting all cigarette advertising on radio or television after January 2, 1971. Congress took this action only against air advertising, and printed media such as newspapers and magazines may accept all such advertising that they care to.

While all this controversy was going on in Washington and in the press, Schenley and McKesson & Robbins bought practically all the time we were offering within 24 hours after the announcement was made. This gave us plenty of publicity and added fuel to the flames. Listeners were writing and telegraphing to us endorsing our stand; only a few opposed it. Some people, generally identifiable as "dry" agitators, told us in no uncertain terms that we were on the road to Hell and were dragging our audience with us.

By now, the Distilled Spirits Institute, which is the spokesman for the industry and works to head off anti-liquor legislation, got into the act. The Institute was worried and warned its members accordingly. But the two companies which had given us orders were not members of this organization so they were free to do what they wanted. When Washington began to talk about laws and other restrictions, they too began to be cautious.

The next pressure put upon WQXR came from Representative Orin Harris, chairman of the Interstate and Foreign Commerce Committee of the House, who asked us about our plans and policy with the inference that the Committee might consider preventive legislation. A similar letter was written to the National Association of Broadcasters. We replied in detail and answered all his questions and did not retreat from our original position. The NAB replied also, but its approach was that it was opposed to such advertising, and it was sure that legislation was not required.

With all this controversy going on, the liquor sponsors had second thoughts. McKesson & Robbins, which had planned to use the radio time to advertise Scotch, decided to use the programs for the promotion of cordials and imported wines. Schenley decided to use the daily broadcasts to institutionalize the name Schenley by putting public service messages in each broadcast. Both sponsors kept their programs on the air for a year or more but, by that time, the publicity value of the campaign was about exhausted, and they stopped. WQXR has been able to attract other hard liquor advertisers from time to time, but it never developed into a major advertising classification. Sponsors selling wines, cordials and liqueurs continue to advertise on WQXR in substantial volume.

Thus this tempest in a flask petered out, but it reinforced our attitude toward broadcasting—that a good station must insist on its own policies if it believes in them. There were post-mortems, of course, and some reflected that we may have made some progress. An editorial in *Broadcasting*, the main trade paper of the industry, said in part, ". . . broadcasters ought not to accept as final that liquor under no circumstances should go on the air. We incline toward the view . . . that the whole subject should be studied with a view toward possible relaxation of the code's prohibition against liquor advertising. This is an appropriate, indeed essential job, for the NAB. Among too many broadcasters the only thing quicker than liquor is the leap at a conclusion. The whole subject deserves sober thought."

Another "touchy" subject among broadcasters is religion and religious broadcasts. The policies of the stations vary widely depending upon the nature of its rural or urban audience, or the geographic location of the station. In some areas, aside from formal religious programs or services, a great amount of gospel music is aired. There are so many religious sects operating via radio in various parts of the country that some stations make a very good living selling sponsored religious broadcasts which use the air waves to raise money.

The Federal Communications Commission has occasionally taken into consideration the amount of time devoted to religious programs in acting upon applications for new or renewal licenses. Too little or too much religion can either be a favorable or unfavorable factor. In the early days of broadcasting, a number of religious groups obtained licenses, and today there are still some which own and operate their own stations.

There is always a demand from groups who want to put religious programs on radio, preferably free, but in many cases on a commercial basis. WQXR had to decide quite early in its career what policy it would pursue. We wanted some, but in view of our emphasis on good music, we did not think it advisable to have very much religious programming. The first condition we made was that we would accept programs only on a commercial basis and these broadcasts would not be allowed to solicit funds over the air. This position avoided the possible accusation that we were giving more free time to one group than to another. An even more important decision was that we would accept a religious broadcast only if it originated from a house of worship at the time of an actual service. This prevented our being forced to take "canned" programs offered to stations by so many sects. We also decided that we would broadcast free any special service from a recognized church if it was a special event of general public interest. This we did from time to time, such as a Christmas Eve mass from St. Patrick's Cathedral in New York or important memorial services from various churches.

In our first year, the New York Society for Ethical Culture offered to buy time for the broadcast of its Sunday morning lectures from their meeting house, and we agreed. The Society still continues these broadcasts after more than 30 years. Other Sundays were taken by the Christian Science Church with services originating from a different church each month in the New York City area. After many years, these broadcasts were no longer wanted by the Christian Science Church and that air time was sold to the Community Church of New York.

Another long-established religious service on WQXR originates from Temple Emanu-El in New York, the largest Reform Jewish congregation in the United States. For many years, it has sponsored the 5:30 to 6 P.M. time for part of its Friday Sabbath service. From time to time we have also had other church sponsorship of actual vesper services from Grace Church, from the Unitarian Church and from famous Trinity Church at Wall Street and Broadway for a weekly noonday worship.

Politics on the air is another "headache" which broadcasters develop with particular severity every two years. The FCC has established specific rules for the handling of political campaigning on the airwaves but, within those rules, there is some latitude for station decision. WQXR made its own regulations very early, and these have proved fair and workable. WQXR will accept political talks only within a reasonable time prior to an election. It sells time to recognized political candidates or to their representatives and does not give free time to any of them unless it is our own station-organized forum where various sides of an issue are debated. It will sell, and has sold, time to candidates on the "fringe" of the extreme right or left provided the representatives are speaking for a recognized party and for a candidate or issue which will appear on the ballot. Spot announcements for candidates or issues are also acceptable provided they are not dramatized to the extent that they mislead the listener. This policy seems to have worked very well and, at election time, WQXR is frequently chosen by candidates and parties because of its reputation for having an influential audience. But even this policy causes problems sometimes. In the fall of 1967, in New York State, one of the important issues before the electorate was the repeal of the so-called "Blaine Amendment" which prohibited the use of public funds toward the support of schools operated by religious denominations. Those in favor of repeal had large financial resources and bought a great many spot announcements on the station, but the opposition had very little money. Apparently our audience for the most part was against the use of state money to support religiously-operated schools. This was evident from the number of protests we received. There was nothing we could do about it beyond exercising our prerogative of not permitting domination of the station. This was the same principle we used in all elections in order to prevent the party or candidate with the most money from buying what we considered an excessive amount of time. For the benefit of those who do not remember the outcome of the election, the repeal was defeated.

Because radio stations operate under a license granted by the Federal Government, at one time it was thought improper for a station to editorialize as a newspaper does. In recent years, however, the FCC has urged stations to editorialize and to take a stand in its community rather than to ignore the issues which are important to its listeners.

We asked the *Times* if we could broadcast one or more editorials the night before they were to be printed in the paper. At first, the management of the *Times* objected to this as anticipating what people could read the next morning, but WQXR's position was backed up by John B. Oakes, the editor of the editorial page, who realized that this was one more way of extending the influence of the editorials while at the same time helping WQXR to meet its public service obligations. For a long time each evening at 6:30, an editorial, selected by the editors, was read on WQXR either by Mr. Oakes or some other member of the editorial board of the *Times* and, occasionally, by an announcer. These editorials dealt with local issues more often than national although important national problems were frequently chosen. Editorials endorsing or supporting political candidates were not broadcast because the FCC "Fairness Doctrine" would require us to notify every other candidate and give him equal time to reply, and this, in turn, would conflict with our overall policy on political campaigning. After every editorial, regardless of its content, an announcer stated that anyone wishing to comment on the editorial might write to the station and the letter would be read on the air if, in the judgment of the station, it was appropriate. We received letters from time to time, but I have always been surprised that we did not get more. I doubt that everyone agreed with the *Times*' editorial point of view to that extent.

Later the FCC frowned upon the policy of a newspaper-owned station using the same editorials as its parent, so this form of editorialization was discontinued. In its stead, at the end of each "Insight" program, Clifton Daniel broadcast what he called "The Last Word" which was in effect an independent editorial comment written by Daniel.

These are the principal matters that are the basic policies that have guided our operations aside from the fundamental ones on advertising and program quality which I have previously discussed in detail.

16

The WQXR Network

As PEOPLE HEARD about WQXR in cities throughout the country, we began to receive inquiries on how to start a station such as ours in other communities. Many well-meaning people thought that all you had to do was to get a license from the FCC, buy some records and put them on the air.

We were anxious to help, for we knew that the more good music stations in operation, the larger the audience would become and that, in turn, would make advertisers and agencies realize more rapidly the potential of good music as a medium. We had to point out to inquirers, however, that they could not hope to succeed unless the audience for good music in their area was large enough to comprise a market worth cultivating by sponsors, unless the station owner was willing to subsidize the operation for its cultural value. Our studies had shown that in large cities such as New York, San Francisco, Boston, Los Angeles, Chicago and Washington, D.C., a maximum of 20 per cent of the total number of radio homes was the potential audience for good music. In other cities, the percentage was lower. It then became a matter of simple arithmetic to determine whether this 20 per cent (or less) constituted a viable market for the advertiser. For example, if a station covers an area containing 4,000,000 radio homes, the potential good music homes would be about 800,000; if there are 1,000,000 homes, you still have a worthwhile market of 200,000, but if your community is 250,000 this leaves only a maximum of 50,000 homes which might have people in them interested in good music.

A number of projects were started and, frequently, the owner would come to New York to talk to us and see how WQXR operated. We were glad to give all the help we could, pointing out the pitfalls as well as the advantages of this kind of programming. Many of these people had hazy ideas about what they planned to do. Some felt that you could accept any kind of advertising interspersed with good music and get the same quality audience we had. Others defined good music as the sort you used to hear in restaurants with a four-piece ensemble playing behind a screen of potted palms. Some went to the other extreme and wanted to program only "way out" experimental compositions as their principal fare. We tried to steer them down a path which would lead to a reasonable chance of success.

By the start of the 1950's, there were some good music stations in successful operation in a few large metropolitan areas. We were frequently exchanging ideas with those in Los Angeles, Chicago, Boston and Washington. This was the era when FM was being introduced, and there seemed to be a special affinity between good music and this improved method of transmission. This tempted many adventurous spirits to enter the field but, unfortunately, many of them had more enthusiasm than capital and failed.

A young man named Ray Kohn had started an FM station in Allentown, Pennsylvania. He had developed a good program and had worked long hours, seven days a week, to keep his head above water. He came to see me and asked whether there was some way we could help him, and he suggested that if he could rebroadcast our programs, that might be his salvation.

It was out of the question for him to bear the cost of a direct telephone line from New York to Allentown, and I suggested that we might experiment to see if he could pick up our FM signal directly over the air and rebroadcast it. Shortly thereafter, by improving his receiving antenna, he was able to get fairly reliable service, and we gave him permission to rebroadcast WQXR for as many hours a day as he wanted provided he also broadcast the New York *Times* news on the hour. This he did and thus WFMZ-FM, Allentown became our first "network" experiment. This operation was not on the air for more than a few days before WFMZ began to get enthusiastic letters from people in the Allentown-Bethlehem-Easton area welcoming the WQXR service. Kohn was building more audience but, unfortunately, this was not bringing him more income. Local advertisers were not convinced that WFMZ had enough audience to make it worth their while.

While this test with Allentown was underway, WQXR received other requests for a similar service. In upstate New York, there was a network of FM stations known as the Rural Radio Network which was owned and operated by the Grange League Federation, a large agricultural cooperative which had started the Rural to transmit special information programs to the farmer members in New York State. The cost of operating this group of stations was large even for as prosperous an organization as G.L.F. so they approached us to see whether, by using some of their hours on the air for our programs, they might attract advertising and make their network profitable or at least self-sustaining.

The nearest G.L.F. affiliate was in Poughkeepsie, New York, and the plan was to pick up the WQXR programs at

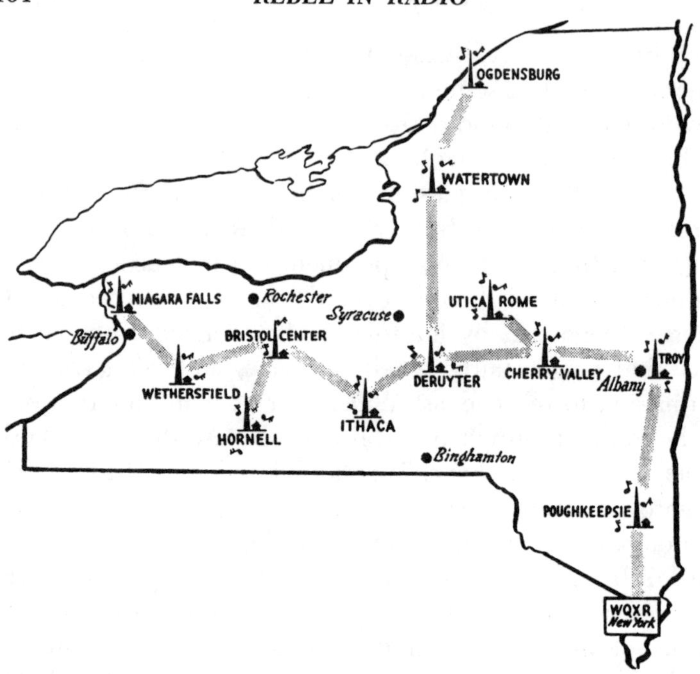

Map showing the transmission route of the WQXR Network in New York State.

that point and relay the signal to the Troy station and thence right across the state to Niagara Falls. We started to make tests. After some stations along the route improved their antennas, the FM relay worked well enough to warrant our hope that a communications system of this kind could operate without telephone wires. We started the service July 1, 1950 with the network taking our evening programs on weekdays and all day on Sunday.

There was immediate action. By and large, upper New York State had very little opportunity to hear any good music on the radio, and many people were hungry for it. Those who first heard the programs told their friends about this newcomer to the air-waves, and soon WQXR via the Rural Radio Network had established a fine audience, not great in size as radio audiences go, but a most valuable one. At no previous time had WQXR received so rapid a response to its programs.

By the time we started our feeds to the Rural, another station, WBIB, in New Haven, Connecticut had joined our network so that on July 1, WQXR programs could be heard from 12 transmitters stretching across New York State to Lake Ontario and into Canada and south into eastern Pennsylvania—all this in addition to the populous Metropolitan Area of New York served by WQXR-AM and FM. In the network areas, one had to have an FM receiver which increased the demand for FM sets. But most manufacturers still underestimated the demand, and good FM sets were hard to find. This slowed the growth of the network for several years until set production increased.

In addition to the excitement over the musical programs, there was great interest in the New York *Times* news broadcasts because they were much more factual and informative than the news they were accustomed to, even though the news periods contained some New York City local news in addition to national and international events. In fact, it was the interest in *Times* news that brought us many other requests to join what by then was beginning to be known as the WQXR Network. Finally, we dropped the Rural identification in favor of WQXR Network in 1953.

Among the early requests for service from WQXR was from a group of FM stations which wanted to relay from New York to Bridgeton, New Jersey, thence to Philadelphia, Baltimore and Washington. Testing showed that the relay was possible only if we had some telephone line links part of the distance to insure a constant signal. We went ahead with this group, but the added cost of lines made it necessary to make important network sales.

These stations, with some additions and subtractions, were the basic WQXR Network for many years. But the operation was never a complete success. There were two sources of trouble: first, the engineering and second, the economic.

The plan to relay programs from station to station without the use of expensive telephone lines was theoretically workable. The sound quality was excellent and superior to the fidelity which could be obtained by any except the most expensive land lines. But this was chain broadcasting in the literal sense and, like all chains, it was no stronger than its weakest link. When all the stations were in good operating condition, everything went smoothly. But if any station along the line broke down, every other station depending on that signal was out of luck. The breakdown hazard was increased because some transmitters were automated, and on high elevations which during emergencies such as heavy snowfalls, could not be reached for maintenance or repairs. Also, some of the stations were trying to make both ends meet and had little money for improvements or stand-by equipment. This same situation caused a deterioration of sound quality when certain stations did not maintain the high-fidelity standards of their transmissions.

WQXR could never tell whether a program was going through properly to the entire network. If a particular station felt impelled to broadcast a local high school basketball game, there was no way for the program to skip a transmitter and, as a result, a part of the network was cut off from the scheduled program, and listeners were disappointed time and time again. These were all engineering hurdles which could have been jumped if the individual stations had had money to modify the equipment or afford personnel to man it properly. They did not, and so these mechanical failures continued to plague us.

Despite such physical obstacles, the WQXR sales force worked hard to sell time for the network and, over the years, they made some substantial sales to sponsors who, in turn, found the advertising productive for them. Later, we found it desirable to have a separate network sales and operating unit in the WQXR organization, and this we established under the management of James Sondheim, an experienced man. We realized that this was going to add to our operating costs, but we felt it was the only method which would determine whether the network could be operated as a profitable extension of the WQXR idea.

By 1963, the WQXR Network comprised 14 stations covering most of New York State, Baltimore, Boston, the Delaware Valley area, Hartford, Philadelphia, Providence, Worcester and Washington, D.C.—one of the most densely populated parts of the United States and one containing a large percentage of people who were keen for our kind of programming.

There was increasing pressure by most of the affiliated stations for more income and, although we were getting some business for them, it was not enough. This forced them to accept programs from local sponsors who did not want good music. Soon some of the individual stations had an image which certainly was not the WQXR image, and thus drove away our faithful listeners and did not attract enough others. In trying to be all things to all men, some of the stations lost their identity as quality broadcasters which added to the difficulties of our network sales department.

This was a dilemma which forced us to review the whole network picture and evaluate its present status and future potential. There was no question that if WQXR programs could be heard over a larger area, there would be a demand for them from an important if not a mass audience. The density of that group varied with the size and nature of the diverse communities being served. Thus far, we had not been able to attract enough national or regional advertisers to support the operation either for WQXR or for the individually owned stations. That was not our greatest worry because we had experienced the same situation in the beginning years of WQXR and had lived through those parlous days. The comparison ends there, however, because we believed in those pioneer days that by continuing our program policy, we would eventually come out on top. We also knew that we had complete control of that policy and could give it every chance to succeed. But the individual network affiliates, forced by lack of money, had to accept almost any kind of program and advertising offered to them, and that spelled the end of quality and cultural appeal which was characteristic of WQXR. The *Times* and WQXR would have been willing to continue the network arrangements a year or two longer to give the plan every chance but, with no control of the affiliate station's programs, we saw the handwriting on the wall and decided to quit. In the early part of 1963, we decided to phase out the network operation and, by that autumn, the WQXR Network was a thing of the past.

This was a great disappointment to us just as it was to those in the northeastern states who were still intermittent listeners to the few WQXR features which were still aired on the network to the bitter end. The network was a "noble experiment."

17

TV and the Future of WQXR

It is not often that a relatively young industry finds itself confronted by a technological development which threatens to put it out of business. It has happened to older industries as when the automobile supplanted the horse and buggy, and we have seen the airplane put the ocean liner in the shade and eliminate almost all long-distance passenger railroad patronage.

But the radio broadcasting business was only about 20 years old when along came TV, and many people who should have known better were ready to attend its funeral services. Anyone who in 1948 had predicted that more than 20 years later local radio still would be alive and thriving financially as never before, would have been regarded as a fit patient for a mental institution. As a matter of fact, the funeral almost did take place, but the resourcefulness of the industry pulled it back from the edge of the grave.

In 1949, television already was showing its future possibilities. There were 50 TV stations on the air, having increased from seven stations two years earlier. In July of that year, *Variety*, the weekly of the entertainment industry, asked me to write an article about what I foresaw for the radio industry. That piece, I believe, best sums up what my thoughts were at that time, so I am quoting it in part:

> If you own a radio station and harken to some of the "experts," the smart thing to do is to have your disk jockey put on a record of a swan song, lock up your transmitter and get into the television business. Maybe these prophets are right, but I don't think they are. Radio is bound to be affected by television, but it is not going to be a "has been." In predicting the future of radio, too many seers think about network radio and forget about the hundreds of successful independent stations throughout the country which have served their communities well and are going to keep on doing it as no television operation can.
>
> Because of the cost of television it must reach a maximum audience in order to pay off. Television, to pay its keep, will have to sell a vast number of people at a low per-capita cost. It necessarily will become a great mass medium, catering to the broadest possible market. Television must aim at a fairly low common denominator of cultural interests.
>
> It's the belief at WQXR that television is no substitute for good music. The recent studies made by WOR seem to substantiate that opinion. Surveys made by The Pulse show that in television homes, good music and other specialized programming do best in competition.
>
> It is not necessary to see an orchestra or artist to enjoy the music. In fact the contrary is true. Those programs of good music which already have been televised have been interesting, but not from the musical standpoint. To watch Toscanini and his orchestra on a television screen is a dramatic rather than a musical experience. Opera and symphonic programs on television will not be a daily occurrence, and music lovers would not want them because watching the television image often is distracting and lessens the enjoyment of the music. The same is true for dance music. It doesn't make much sense to stare at a television screen for hours just because you like to listen to dance music. And if you really want to dance, keeping your eye on the visual image is hard on the neck!
>
> This would seem to indicate that the future of radio lies in specialization. There is little likelihood that television will be

> able to specialize the way radio can. And there are many fields of specialization in radio: local sports, hot jazz, spot news, special events, programs specifically directed to large groups such as organized labor and foreign language segments of our population.
>
> The great danger which the independent station operator faces at this time is the temptation to pull in his horns, lower the standards of his programs and advertising, economize at the expense of holding his audience and then blame everything on television.
>
> Nothing will stop television. It is here and is bound to become a most influential mass medium. But there is a way to keep a successful radio station successful and that is to improve its programs, increase its service to the public and concentrate on the thing it does best and which it can do more economically than television can. By building rather than by tearing down, radio specialists will hold and actually increase their audiences.

I must admit that although I sincerely believed in what I wrote in 1949, and I said the same thing in talks before various broadcast and advertising groups, some of my thinking may have been biased and perhaps I was, to some extent, whistling in the darkness of the threatening television eclipse. When I made these comments on the future of radio, there were only 50 TV stations on the air. In June 1972, there were 774 commercial stations authorized by the FCC and 701 actually broadcasting. Radio stations have also multiplied. In 1949, there were about 2,600 AM and FM commercial stations, now there are more than 6,700. Several factors which none of us contemplated back then are: the growth in the number of FM broadcasters, the enormous demand for small inexpensive receivers and the universality of automobile listening—all of which have made radio programs available to millions of people no matter where they may be.

Our predictions about the future of TV and what we thought it would become were the result of our special experience with WQXR. The *Times* was not blind to the possibilities of TV as a competitor in the field of communications, and Arthur Sulzberger asked me to study the problem with our technical people and prepare a recommendation for the *Times.* Meanwhile, we actually began to prepare a TV application to the Federal Communications Commission.

The report contained the generalizations which I have already mentioned and then went on to question the kind of TV station the *Times* would want to operate. We pointed out that if the *Times* wanted on TV the quality it had on WQXR, the station could not be self-supporting and would, in fact, be a very costly luxury for the *Times* to support. On the other hand, the report pointed out that television was going to be a mass entertainment medium and, as such, might be a very profitable investment for the paper. To embark on television, we pointed out, was a decision that Sulzberger alone would have to make.

A few days after he received our written report, he called me on the phone and said, "The *Times* is not in the entertainment business—do not file the application for a station." The correctness of the decision, granted that the *Times* did not want TV just for a speculative investment, is to be found in the inability of cultural, educational and quality TV operations such as National Educational Television to exist without heavy subsidies from the government, wealthy individuals and foundations. Television is essentially an entertainment medium aimed at the mass audience, with a smattering of "specials" which are generally at a higher level of entertainment or information. Meanwhile, "back at the ranch," WQXR has kept pretty much to the plans made in 1936 and still offers small but effective competition to the TV stations.

But what about the future?

In my opinion, there will always be a need for superior radio programming. The need will increase because there are going to be more and more better-educated people who will not be satisfied with constant lowering of standards by the run-of-the-mill radio and television programming. When all is said and done, though, the economic viability of a station such as WQXR will be the determining factor.

The most severe blow to the ability of WQXR to pay its way was delivered by the United States Government via the Federal Communications Commission in 1966. A few years before, the FCC had announced a proposed rule which provided that in cities of 100,000 population or more, an FM station could duplicate the programs of its AM affiliate no more than 50 per cent of the time. Heretofore, there had been no restriction on total duplication. The proposed rule had been advocated by the "liberal" members of the Commission and was voted in by a narrow margin. The argument for the change was that it would give cities a greater variety of programs for the public to choose from.

One can understand why, in a small community with perhaps one or two licensees, the splitting of AM and FM might create needed additional types of programming. But in large cities of 100,000 or more, there already was a wide choice of radio fare. For instance, in New York City there were at least 40 stations available to the listener, including both AM and FM. Among this multiplicity of stations, there was every kind of broadcast that the most experienced radio men had been able to conceive. So what more could the New York populace want? Another ridiculous provision of the proposed rule was that a station could comply if a program was repeated after a lapse of 24 hours. To my mind, this negation of the expressed purpose of getting more variety could only mean that the eventual goal of the FCC's separation edict was to force licensees to dispose of either the AM or FM side of their operation.

WQXR, as well as some other stations, filed protests with the Commission, but many station owners did not realize until it was too late that separation was one of the worst things that could happen to them. The owners of struggling FM-only stations backed the proposal in the mistaken belief that it would help them competitively, not realizing, in many cases, that the change would vastly increase the number of FM-only stations and thus increase, rather than decrease, the intensity of the battle for sponsor support.

For several years, the FCC accepted the briefs filed by those of us who were in the opposition and, from time to time, delayed the date of the effectiveness of the rule. This suggested that there were some members of the FCC who were having second thoughts. But they did not prevail and, finally, the date for the changeover was fixed for no later than January 1, 1967.

It gave WQXR about six months to plan and prepare for the separation, and it was not a simple task. First, the actual physical changes which would be necessary had to be planned. This involved the construction of three new studios, an additional control room and modifications of our whole broadcast set-up so that we could operate as two separate stations simultaneously. This was an expensive undertaking and, of course, the construction work had to be done without interfering with our regular routine. This was our first and least expensive "headache."

The chief task was to decide how the programs would differ during the hours that the AM and FM would be operating independently. This involved giving consideration to many variations, even including rock n' roll which was quickly dismissed. An all-news operation seemed attractive, but when we analyzed the costs involved in running a news station which would be in character with the *Times,* we found it economically impossible. Music being our forte, it was decided that the new program had to be in the realm of music and, finally, we all agreed that it should be "light classical," that in-between realm of tuneful favorites for "middlebrows."

Beginning January 1, things are going to be different at WQXR. Every day between 9 and 6 you'll have two different WQXR's working for you.

WQXR-FM will be strictly classical. Symphonies. Concertos. Operas. Chamber Works. And no chatter. Only the hourly New York Times News.

WQXR-AM will have a lighter tone. Light classics predominantly. Easy-listening music. Broadway shows. Jazz and Folk. With sophisticated personality hosts. Times news on the hour and extra half-hourly news roundups.

Two different WQXR's. Two different moods. Both with the same standard of excellence.

We'd like you to take sides.

Both sides—WQXR-AM and WQXR-FM.

96.3 classical FM WQXR 1560 lighter AM

Announcement of the ill-fated separation of AM and FM programming.

The plan seemed to work out best by duplicating the programs from 6 A.M. to 9 A.M. and then separate until 6 P.M., when the two stations would be united until midnight, with the hourly *New York Times* news heard on both stations at all times. This is an approximation of the schedule, and it met the new FCC regulations.

Next we had to decide whether the AM or the FM would carry the new programming. We decided to put it on AM and to move all "talk programs" to the AM, leaving the FM for the best music. Our thinking was logical. Serious music is entitled to faithful reproduction, and FM plus stereo could certainly do that better than AM. I believe we made the wrong decision, even though all the logic favored it, for it meant that our listeners could not hear any really fine music during the day on their AM sets. We were abandoning the good music field on AM in the daytime except for the occasional symphonic programs on WNYC. We also realized that the automobile listener would have to listen to our lighter music, for very few cars had FM receivers. In fact, we even thought that the car listener might prefer light classical sounds as he drove from place to place. But we were very, very wrong.

On the other hand, had we put the light program on FM, I am sure that we would have had equal dissatisfaction. I have gone into the details of this change forced upon us and our listeners by the bureaucratic insistence of the FCC to show, once again, how impossible it is to satisfy all listeners all the time—and how much hardship this bureaucratic ruling brought, not only to WQXR, but to the tens of thousands of listeners who resented the change and who blamed us, and could not understand that we were not free agents.

In the hope of holding our own, and perhaps gaining a new audience, we decided on a promotion and advertising campaign to cost $100,000 to accomplish this for us. This campaign was to be concentrated in about three months. When launched, it attracted people to the new daytime music but, when they listened, they did not like it and flooded us with protests. Even Madison Avenue did not take to it so we did not attract any new business and, in fact, it harmed the WQXR commercial image greatly.

When we saw what was happening, something had to be done about it. We had made a mistake in under-estimating the musical taste of our established audience and had over-estimated the musical taste of the people we had hoped to attract. The separation started on January 1, 1967, and we made up our minds that a change must be made before July 31 in order to start the fall season on the right track.

Gradually all the light program was dropped, and AM got the same kind of programs as FM, some of the hours being repeats of what had been heard on FM the previous week. As a result, we had two good music stations giving the WQXR public what it liked. But it could have had the same music without the separation just as it had before the edict, and without the expense to us of duplicating a large part of our engineering, announcing and programming staffs. To my mind, this example of government regulation covering a whole class of businesses without due consideration of the special problems involved in individual cases is wasteful and not in the public interest. If a station such as WQXR, which for more than 30 years had been trying to raise the standards of radio, was not recognized as a special case, what inducement was it to WQXR or to any other broadcaster to try to help people who want more intelligence and culture on the air?

Of course WQXR made further efforts to have the rule rescinded or modified. These approaches to the Federal Communications Commission were met with little flexibility. Whereas the FCC admitted that our situation programwise and our specialized audience was different, its argument was that the Commission could not make an exception for WQXR. When we consulted our Washington attorneys, they advised that legal steps would not force the Commission to change the rule.

Meanwhile the FCC made no attempt, as far as we know, to inquire from the New York audience whether the non-duplication rule was of any advantage to the average WQXR listener.

This situation continued for five years, adding to the deficit operation of WQXR which in this period also was suffering from a decreasing volume of advertising and greatly increased costs of operation. As a consequence, in 1971 the *New York Times* decided to sell the station and it was announced in a story in the *Times* that WQXR was for sale. This caused consternation among many WQXR listeners who anticipated a new ownership which would certainly not maintain the good music policy which many of them had enjoyed for 36 years.

The same economic pressure was driving other good music stations out of the field. One which had a fine reputation was WGMS in Washington, D.C. which WQXR had help get started many years before. WGMS announced that because of its inability to support its good-music policy as an AM-FM separation operation, it was going to change to a mass-appeal type of station. Hardly had the announcement been made than protests came from listeners. In politically sensitive Washington this began to have its effect. The Washington audience of WGMS was evidently the same kind that WQXR had in the New York metropolitan area—people who were leaders in business, the professions, members of Congress, important people in education and in Washington society—people who were leaders and who influenced others. Before long, groups formed to oppose the proposed change and when they discovered that the 50 per cent non-duplication rule was partly responsible for WGMS's financial problem, they brought pressure on the station's ownership (RKO General, Inc.) to apply to the FCC for 100 per cent duplication of AM on its FM outlet and continue its good music policy.

This the owners agreed to do and the Commission, mindful of the strong public backing and its political effects, gave WGMS permission in March 1972 to duplicate.

The effect on WQXR was immediate. The *Times* announced on March 29, 1972 that the station was no longer for sale and that it would apply at once to obtain the same waiver which WGMS had won.

It took until near the end of June for the FCC to act, and on June 30, 1972 WQXR announced that the waiver of the rule had been granted and that duplication would start on July 9th. It was also stated that, because of the elimination of the 50% rule, the station would be able to broadcast more symphonic music than it had in the past few years. The story in the *Times* reporting this news disclosed for the first time to the public that WQXR had operated at a loss of $300,000 for the year 1971.

This development hopefully will enable the station to continue along the same program path which it blazed in 1936 and which in the last few years had been endangered.

There is also another threat to the future of WQXR, again from government sources. As I have mentioned, the separation of AM and FM programming has been interpreted by some as the first step in eventually prohibiting a broadcaster from owning more than one kind of station in a market. If he owned an AM, he could not own an FM or a TV station, and vice versa.

In 1968, the FCC made it clear that in considering future applications for licenses, it would bar multiple ownership in a single market to protect the public against a monopoly of communications. This rule was to apply to new applications. But in the summer of 1968, the Department of Justice sent a memorandum to the FCC suggesting that it consider the advisability of reviewing all multiple ownership situations regardless of the length of time they had existed and eliminating the ownership of broadcasting stations by newspapers, other publications or informational channels, presumably including cable television (CATV).

This policy has not been enforced in most instances but it is a threat as long as the Department of Justice is thinking along those lines. If the rule ever went into effect, all that the *New York Times* has spent in effort and money to maintain WQXR as a better radio service would come to nought. In my opinion, the chance of the rule applying to established stations is a long time off. It will be opposed by important communications interests. It should also be remembered that many Senators and Representatives have financial interests in both broadcasting and publishing. But it could happen and, if it does and the *Times* is forced to divest itself of WQXR, the future of the station as a leader in intelligent broadcasting is dark indeed.

However, planning the future of WQXR should not be handicapped by fear of government restraints or shortsighted FCC policy. Its future depends upon the steadfast determination of the *Times* and the present generation of *Times* and WQXR management, headed by Arthur Ochs ("Punch") Sulzberger, to maintain the aims which have guided the station in the past. It is my hope that the principles laid down by Jack Hogan and me in 1936 will always be kept in mind, despite many temptations to yield to a more profitable format. This does not mean that the management should be blind and deaf to what is going on in the world. Who could be in this epoch of social revolution? But no matter what goes on, there are such things as intelligence, culture and good taste. There will be more people in the years to come who will desire what these three things stand for, and WQXR will, I hope, be there to give them what they want.

"He dreamed 'of
some world far
from ours,
where music
and moonlight
and feeling
are one.'"

Shelley

Index